THE
WINE-DARK
SEA WITHIN

THE
WINE-DARK
SEA WITHIN

A Turbulent History of Blood

DHUN H. SETHNA

BASIC BOOKS
New York

Cover design by Rebecca Lown
Cover images © Valentin Agapov / Shutterstock.com; © YASNARADA / Shutterstock.com; © Lightkite / Shutterstock.com

Basic Books
Hachette Book Group
1290 Avenue of the Americas, New York, NY 10104
www.basicbooks.com

Printed in the United States of America
First Edition: June 2022

Published by Basic Books, an imprint of Perseus Books, LLC, a subsidiary of Hachette Book Group, Inc. The Basic Books name and logo is a trademark of the Hachette Book Group.

The Hachette Speakers Bureau provides a wide range of authors for speaking events. To find out more, go to www.hachettespeakersbureau.com or call (866) 376-6591.

The publisher is not responsible for websites (or their content) that are not owned by the publisher.

Print book interior design by Linda Mark

Library of Congress Cataloging-in-Publication Data
Names: Sethna, Dhun H., author.
Title: The wine-dark sea within: a turbulent history of blood / Dhun H. Sethna.
Description: First edition. | New York: Basic Books, 2022. | Includes bibliographical references and index.
Identifiers: LCCN 2021061890 | ISBN 9781541600669 (hardcover) | ISBN 9781541600676 (ebook)
Subjects: LCSH: Blood—Circulation—History. | Medicine—History—Popular works.
Classification: LCC QP101.4 .S46 2022 | DDC 612.1/3—dc23/eng/20220211
LC record available at https://lccn.loc.gov/2021061890

ISBNs: 9781541600669 (hardcover), 9781541600676 (ebook)

LSC-C

Printing 1, 2022

For
Avi
my bride
who still abides with me

Contents

I stand in awe of my body, this matter to which I am bound.

— HENRY DAVID THOREAU

Introduction

THIS BOOK IS THE BIOGRAPHY OF AN IDEA, THE IDEA THAT BLOOD circulates around the body. It may seem a common, even obvious notion. Yet that familiar concept that the heart is an organ that pumps blood and oxygen through the arteries, with the "waste" returning by way of the veins, took over two thousand years to develop. When it was established, it revolutionized the life sciences and inaugurated modern medicine. In importance, it stands alongside the Aristotelian Corpus that laid the foundation for the biological sciences, and Darwin's theory of natural selection. And like those ideas, its development was largely a solitary effort, conceived, begun, and completed by a single individual, the English physician William Harvey. He published his discovery in 1628 as a slim volume titled *Exercitatio anatomica de motu cordis et sanguinius in animalibus* (*Anatomical Exercise on the Motion of the Heart and Blood in Animals*), abbreviated to *De motu cordis*. Through it, Harvey lived not only one of the greatest adventures of all time in medicine but, in the process, experienced the insecurity, vulnerability, and frailty of the human condition. He is a man of the present; he belongs everywhere.

Blood circulation, as is now understood, is a double system. There is a circuit through the body (the systemic circulation) as well as one through the lungs (the pulmonary circulation). Each circuit is spoken of as a circulation because the circle is a symbol that ends at the point at which it begins. The historical unraveling of each circulation is a theme of this book. The heart, too, is in reality an assembly of two hearts that work in harmony at two different tasks. The right heart chamber propels blood to the lungs; the left chamber distributes it to the other organs and limbs. Because the motion of life-preserving blood was realized to be closely allied to the breath of life and the maintenance of a constant body temperature, those three processes merged to lie at the very core of the new physiology. Hence, the province of this narrative is also the development of a theory of animal heat and the early physiology of respiration.

THE DISCOVERY OF THE CIRCULATION WAS A GAME CHANGER IN THE history of the life sciences. It ushered in a new quantitative way of thinking that spawned further innovations in disease management without which medicine, as we know it, would have been impossible. Harvey's hydraulic description of the circulating blood, founded on pumps and pipes, laid the groundwork for a quantifiable, mechanical system of cardiovascular physiology that led to our modern quantitative way of thinking in terms of blood velocity, vascular resistance, blood pressure, pulse waves, and so on, as well as their quantitative changes under varying pathophysiological conditions, and the effects of abnormal velocities and pressures on the body organs. If blood circulated, then new questions needed answers. What was the need for blood to perpetually go round in a circle? What did it carry when flowing in such a fashion and why? How and where did it take up its stuff? How, where, and why did it part with it? Those answers unraveled a comprehensible picture of the working of the human organism and established a physiological basis for modern medical practice.

It followed that mechanisms of disease were amended and expanded. A circulation throughout the body meant that diseases could result not only from imbalances of internal "humors," as was believed until our mid-nineteenth century, but also by noxious agents from *outside* that could enter the bloodstream and travel to all tissues. As a corollary, some diseases could arise from an "insufficiency" of blood circulation to vital organs because of obstructions within the arterial conduits, including those of the heart and brain, which led to our understanding of how heart attacks and strokes occur. They remain the primary causes of disability and death today.

Aspects of today's therapies, such as intravenous infusions (as in chemotherapy) or subcutaneous injections (like insulin shots), even nasal sprays for allergy, could only have been conceived after it was understood that substances introduced into the bloodstream at one site, or even breathed in, are transported to any and every other site because the blood circulates. Routine interventions such as heart catheterizations and stent placements within arteries, and the flotation of pacemaker and defibrillator electrodes through veins, all require a unidirectional blood flow within blood vessels into, or from, the heart chambers as described by Harvey. Lifesaving support systems such as dialysis units as well as heart-lung machines that allow "open heart" surgeries are essentially extensions of the concept to extracorporeal circulations, and heart-assist devices (artificial hearts) that save lives during extreme acute illness, or serve as alternatives to heart transplantation, too, rely on a circulation model. A cogent outcome of Harvey's groundbreaking discovery is our present understanding of heart failure, which is our most expensive hospital diagnosis for persons over age sixty-five. Contemporary therapy not only embraces the heart as a failing pump but also addresses the circulating chemical abnormalities that cause the heart muscle to deteriorate.

SCIENTIFIC DISCOVERY IS A COMPLEX PHENOMENON. THE AMERICAN philosopher of science Thomas Kuhn provided what is arguably its best description.[1] The process, according to Kuhn, begins with the recognition of an inconsistency in the normal expectations of things. Next comes an extended exploration of that anomaly, and the progression ends only when the new knowledge itself becomes clear. The normal state of things is now adjusted to accommodate that learning, and what Kuhn calls a "paradigm shift" takes place. Polish philosopher-physician Ludwik Fleck added that, on its way to assimilation and recognition by the "thought-collective," the paradigm "quietly removes its individual idiosyncrasy, shakes it loose from its originating, psychological, and historical contexts, and adapts it to the schemata of the total system."[2] In the process, the system itself is redefined.

The idea of a circulation was one such paradigm shift. In the case of Harvey's circulation, the dominant system, which was the Galenic model that had prevailed undisputed for fifteen centuries, could not be redefined but had to be replaced. The Roman physician Galen had pictured blood as flowing back and forth in the vessels, like a tidal ebb and flow. He had imagined two separate systems of vessels, the veins and the arteries, arising from two different organs, the liver and the heart, which offered blood to all parts of the body. Harvey's revolutionary discovery drove out those obsolete beliefs with new items of fact.

Inherent in the Kuhnian-Fleck model, too, is the notion that discovery in science is a discontinuous process that invariably leads to a sequence of "incommensurable" points of view, influenced by prevailing social and cultural structures, attitudes, and worldviews. The discovery of blood circulation is about just such discontinuous watershed moments in cardiovascular physiology, with every instant saddled by its own baggage of "pre-ideas" that is carried forward. Discovery, thus, is a creative process that takes time. It is a progression to something, but also from something, and the latter something is left behind even though it might have done some good service for a while.

Characteristic, too, in the discovery process is the progress of science by analogy, with certain analogies preventing and others enabling the way to the truth. From earliest times, thinkers have been captivated by analogies to describe the realities of nature. The analogy of blood flow to the ebb and flow of Homer's "wine-dark sea" and Aristotle's comparison of the vascular system to an irrigation canal, as well as the analogy between life and respiration or combustion pervades the entire narrative, from Galen to Boyle. Galen invents his "natural faculty of attraction" in the body from the affinity between a magnet and iron. Empedocles in Sicily draws on the workings of the Egyptian clepsydra (water clock) to enunciate his novel theory of cardiorespiratory physiology, as does Erasistratus at Alexandria from the phenomenon of *horror vacui* (nature abhors a vacuum). Descartes compares fermentation in the heart chamber to log fires. Finally, Harvey seeks comfort in Aristotle's philosophy of circles, and finds confirmation for the heart's function as a mechanical pump in Caus's mechanical fire pump.

Harvey's own work encompassed two simultaneous paradigm shifts: the mechanism of contraction of the heart, and then the circulation of blood. The correct analysis of the heart's motion as a mechanical pump, and only as a pump, that ejected blood into the vessels at each contraction was a central innovation essential to his scheme. Before Harvey, the accepted process of heart function, which came from Aristotle, was a heat-driven "fermentation" of blood within the heart that caused that organ to expand and, like "boiling milk spilling over," caused an overflowing of blood into the aorta.

The crowning point of the discovery arrived during the scientific revolution of the seventeenth century, an era of "promise with disappointment, and resilience with despair." Further development of Harvey's ideas linked a galaxy of the greatest minds and some of the oddest personalities in British science—John Locke, Christopher Wren, Robert Hooke, Henry Cavendish, Joseph Priestley, and their peers; the Scot Joseph Black; the Anglo-Irish "Skeptical Chymist" Robert Boyle with the Oxford Chemists, as well as the French Europeans René Descartes and Antoine Lavoisier. Together, they dissolved the misconceptions of two thousand years of physiology. They, in turn, stood upon the shoulders of the now forgotten pioneers of the more ancient Ionian, Athenian, and Alexandrian intellectual revolutions, men like Alcmaeon of Croton, Diogenes of Apollonia, Hippocrates and Praxagoras both of Cos, the Sicilian Empedocles, the Alexandrians Herophilus and Erasistratus, and the Roman Galen, who all broke ground to understand the natural world within us. Plato and Aristotle, too, played their significant parts. And, like everything else, it all began with Homer — with the ebb and flow of his "wine-dark sea."

THIS BOOK IS DIVIDED INTO SEVEN SECTIONS, EACH MARKING A Kuhnian milestone on the road to a circulation. Because a proper appreciation of the idea of a circulation may be reached, as a link in a chain, only after familiarity with antecedent strongly contrasted concepts, the first section, titled "A Wine-Dark Sea Within," opens the narrative with extracts on blood flow drawn from the *Nei ching* (the oldest medical textbook known) in ancient China, the medical *Ebers Papyrus* and the *Edwin Smith Surgical Papyrus* from archaic Egypt, and the *Atharva Veda* of ancient India. The story begins properly in ancient Greece, where prototheories about air, blood, nutrition, body heat, and body waste, as well as heart function, are developed, and the types of conduits that carry them are defined.

Those prototheories are amalgamated and advanced into *systems* of air and blood at Alexandria in Egypt and at Rome in the next section, "Wrinkles in Blood Flow," which rounds out our understanding of the anatomy and physiology of the heart and vascular system as known to the ancient world. The physician Erasistratus at the Alexandrian school applies the physicist Strato's recently announced theory of *horror vacui* — "nature abhors a vacuum" — to enunciate a unique physiology of forward blood flow. Blending and codifying all the prototheories forged by his predecessors, a provincial Greek physician, Galen, practicing in Rome, cobbles together a plausible theoretical scheme of cardiovascular physiology that has no basis in truth, yet becomes the springboard for body function, disease, and treatment for the next fifteen hundred years. The significance of Galen in our story cannot be overestimated. Just as the writings of Aristotle and Plato are the basis for science and philosophy, respectively, so Galen's much more voluminous works become the foundation of medicine and physiology. They are still, by far, the largest surviving oeuvre of any ancient author. His works run to twenty-two volumes, including about 150 titles.

Through fantastic inventiveness and contrivances, Galen creates a marvelous and sophisticated theoretical edifice that appears to explain all aspects of body physiology and becomes accepted as dogma. Because he writes in Greek, his initial influence is in Greek-speaking Alexandria and Byzantium and in the remnants of the Eastern Roman Empire. With the rise of Arabic learning under Islam, his system of the heart and blood flow becomes their canon, and the Arabs make him their model physician. With the *Reconquista* of Muslim Spain by the Christians and the translation of Arabic medical texts into Latin, Galen is resurrected by the European West. Through the rediscovery of classical Greece and Rome in Italy, Galen enters the mainstream of Renaissance medical thought. The enormous influence of Galen's works, which encompass all aspects of physiological theory and medical practice, extends well into our nineteenth century. In fact, it stands as a formidable obstacle that precludes an earlier discovery of a true circulation for over a millennium.

But Galen's invented physiology proves to be too sophisticated! The Renaissance is afoot, and cracks in the old Galenic system are exposed in the next section, "New Hearts for Old," by Leonardo da Vinci, who explores new avenues through mechanical models and experiments to understand blood flow and heart valve function hydraulically. Aspects of Galen's dogma are questioned publicly by Venetian anatomists as well as at Padua by the young Andreas Vesalius in his groundbreaking and eventually influential anatomy book *De humani corporis fabrica libra septem* (*The seven books on man's bodily works*) published in 1543 when he was age twenty-eight. The academic world takes up arms in defense of Galen. In those circles, everyone is pleased to hear the genius of another colleague depreciated. Vesalius is discredited; shunned, he leaves academia.

In his extension of the Kuhnian model, Fleck sometimes compared a scientific discoverer to Christopher Columbus, who sought India but instead found the West Indies, a not uncommon scenario in science as is manifest in the discovery of the pulmonary circulation, described in the subsequent section "Air and Blood." Its knowledge is one of the necessary pieces of evidence that Harvey needs before he can complete the systemic circuit of blood. Because the two circulations are *in series*, they merge into each other to become only *one* circulation. The pulmonary circulation is discovered for a variety of reasons: anatomical, physiological, and even theological. The Unitarian theologian-physician Michael Servetus settles on the lungs as the proper site for the broadest possible contact between air and blood, and thus outlines a bona fide pulmonary circuit whose purpose is to infuse man's Christian soul with God's vital spirit. About the same time, the papal physician Realdo Colombo unequivocally describes the pulmonary circulation based on sound conclusions drawn from vivisection and experiment. But there is reason to believe that he may have been aware of the work of the medieval Arab physician Ibn al-Nafis, who is the first to describe the pulmonary circuit. The controversy and debate in support of and against this active revolt against Galenic doctrine are detailed, and the battle for the matter of priority in this discovery is addressed.

William Harvey's research leading to the publication of *De motu cordis* forms the whole next section, "Physiology of Circles." He forges his weapons by making observations through vivisections and by planning and conducting probatory experiments with ligatures to determine the direction of the flows of blood in the arteries and veins, thereby demonstrating the unidirectional return of blood to the heart through the veins. Using novel quantitative reasoning, he infers the systemic blood circuit. Paradoxically, while he dismisses the widely accepted, but never seen, pores in the heart's septum (partition) invented by Galen, he promotes the existence of an equally unseen component, the capillaries. They are demonstrated soon after Harvey's death by Malpighi using Galileo's newly created microscope. The Harveyan circulation is now completed. Those who understand the circulation recognize its medical potential, and the era of intravenous therapy and rejuvenating blood transfusions is launched.

But old errors die hard! Harvey's discovery becomes one of the thorniest scientific dilemmas of the seventeenth century and the most defiant to acceptance. The next section, "A World on Fire," aptly describes the academic maelstrom that breaks out after *De motu cordis* is published. Human nature is revealed at its best and worst as battle lines are drawn in defense of the Galenic system as well as the new Harveyan physiology in England, Spain and Portugal, France, Germany, Denmark, the Low Countries, and eventually in the New World. René Descartes doggedly opposes the function of the heart as a pump, and a vitriolic attack is launched in Paris against the whole Harveyan concept. Laypersons (John Donne, Samuel Pepys, John Evelyn, Molière, Racine, La Fontaine) are not exempt from the controversies.

After a generation of conflict, the circulation is finally accepted. But the "purpose" of a circulation remains unresolved. A vexing question remains: "What is the one clear end of a circulation?" Furthermore, Harvey has no conception of the mechanism of changes the blood undergoes in the lungs. Light on both issues is shed by the pioneering chemical work of Boyle, Hooke, Lower, and Mayow, who are collectively called the "Oxford Chemists," as described in the final section, "Consider the Air." Robert Boyle, Harvey's young colleague, begins his canonical experiments on combustion and respiration assisted by the other Oxford Chemists, who collectively set into motion an inquiry regarding the composition of air. A new era in physiology and chemistry follows with the isolation of oxygen, hydrogen, and carbon dioxide by Joseph Black, Priestley, Cavendish, and finally Lavoisier, to establish the proper place and role of Harvey's circulation in metabolism, body heat, and life.

AS WITH ALL GREAT DISCOVERIES, THE ROAD TO THE CIRCULATION spawns some of the bitterest scientific disputes. In the inquiring temper of an age when printing presses are swiftly spreading the word of ongoing innovations, there is much to gain materially for the pioneer who first unveils a truth of nature. If the expectation of pecuniary reward is a good incentive, so, too, is the prospect of fame. After all, greed and ambition are still the mainsprings of human conduct that reveal us to ourselves, and to others, by expressing the better angels of our natures and the baser natures of our beings. As significant is the collective crowd psychology of the discovery process. The history of the circulation of blood is ripe with false attributions and contested claims of priority. They add a human dimension to the narrative. This book is a very human story.

A WINE-DARK SEA WITHIN

Achilles, from his friends withdrawing, sat
Beside the hoary ocean-marge, and gazed
Upon the wine dark sea beyond.

—HOMER

one

The Oriental Heritage

Everything must have a beginning . . .
and that beginning must be linked
to something that went before.

—MARY SHELLEY

LONG, LONG AGO, AROUND 18,000 BCE, AN AURIGNACIAN ARTIST drew in red ocher the outline of a mammoth on the wall of a cavern at Pindal in Spain. In the center of its chest, she placed a large dark spot that was the heart. It is our earliest known representation of that organ.[1] She must have known that the spot was life-giving because Paleolithic hunters understood that when it was struck by a spear, the huge creature, spilling its mysterious red fluid, lay down in a deep slumber from which it could not be awakened. Clearly, there was some force that seemed to drive all living things, not only the graceful flight of birds but also the invisible wind and the lightning that leapt from clouds as well as the endless flux and reflux of waves. To be ignorant of motion was to be ignorant of nature. Then, perhaps,

3

she wondered whether there was a living sea within her own body that ebbed and flowed like the tides, which animated her being and prompted all thought.

Analogies between the earth's rivers and the streams of blood within the body have recurred since archaic times. Our oldest surviving medical textbook, the *Nei ching* (also called *Huangdi neijing*), talks of channels within the body in which the blood streams continuously in a circle:

> The harmful effects of wind and rain enter the system first through the skin, being then conveyed to the tiniest vessels. When those are full the blood goes to the [?] and these in turn empty into the large vessels. The blood flows continuously in a circle and never stops.[2]

That work was composed around 2650 BCE by the legendary Chinese emperor Huang Ti, a monarch who allegedly displayed divine intelligence from the moment of his birth. The current version of the text dates from about 300 BCE. The structure of the heart and blood vessels in the *Nei ching* is schematized to match the numerical harmony of the universe. The twelve months are represented by twelve groups of large vessels and twelve vessels of lesser import through which move the Yin and Yang, which Chinese philosophy describes as contrary forces that are complementary and interdependent in the natural world. Four main vessels correspond to the four seasons; the total number of vessels are 364. The heart, representing the Yang, controls the blood and forms the root of life. The text reads: "All the blood is under the jurisdiction of the heart."

A later Hindu Vedic manuscript called the *Atharva Veda*, or *Book of the Knowledge of Magic*, composed in India around 700 BCE, compares the heart to a lotus flower with nine gates. Scholars have speculated that those "gates," using modern terminology, represented the four pulmonary veins, the superior and inferior vena cava, the aorta, and the right and left pulmonary arteries. Verse seventeen of Book I, titled "To Stop the Vessels of the Body," mentions blood-containing "vessels," including a "great tube," distributed throughout the body. Verse two of Book X, titled "The Wonderful Structure of Man," suggests a flow of fluids, like rivers, running in different directions within the human body:

> Who disposed of him waters, moving apart, much moving,
> produced for river running, strong, ruddy, red, dark and
> turbid, upward, downward, crosswise in man?[3]

The *Charaka samhita* (*Charaka's collection*), assembled by a Buddhist physician Charaka in the first century CE, is a collection of medical knowledge from earlier sources, reflecting, for the most part, the teachings of the Ayurvedic School of Atreya that flourished from 600 BCE at the city of Taxila in the Punjab province of Pakistan. Two types of tubular "vessels" are described: the smaller ones are called *siras*, and the larger ones are *dhamanis*.[4] They carry blood, life breath, chyle, urine, and semen. No functional distinction is made between arteries, veins, nerves, or ducts. Division into *dhamanis* and *siras* is based purely on caliber: a *dhamani* is thick walled and the *sira* is thin. The heart, regarded as the seat of consciousness, is the source of ten *dhamanis* that distribute the breath, intelligence, and "vital spirit" (*rasa*) to the body. Blood, carried by the "vessels," is believed to originate in the liver and spleen. One passage from Charaka's text has been especially intriguing to circulation physiologists:

> From that center [? the heart] emanate the vessels carrying blood
> into all parts of the body—the element that nourishes the life of all
> animals and without which life would be extinct. It is that nourish-
> ment that goes to nourish the unborn child in the womb, and which
> after flowing into its body, returns to the mother's heart.[5]

About a century after Charaka, a prominent surgeon, Susruta of Benares, compiled another tome of medical practice called the *Susruta samhita*.[6] Susruta presents an abstract scheme of tubular "vessels" consisting of twenty-four *dhamanis* that carry "chyle" from the heart throughout the body: ten going upward, ten downward, and four horizontally. The *siras*, which are numbered seven hundred, spring from the navel of the belly like the spokes of a wheel. They carry both blood and air. The heart is compared to a lotus bud hanging with its apex downward.

A SIGNIFICANT PORTION OF THE EGYPTIAN MEDICAL COMPILATION called the *Ebers Papyrus* (c. 1550 BCE) is a summary of "vessels" functioning as irrigating canals. That is not surprising considering ancient Egypt's economic dependence on the Nile. They are recorded as "46 vessels go from the heart to every limb" in one section of the text, but only twenty-two vessels are mentioned in another.[7] There is no differentiation of the "vessels" into nerves, ducts, arteries, and veins; the same word *metu* is used interchangeably. Some vessels bring air to the heart and other parts of the body; the remainder carry blood, urine, sperm, and saliva. The anatomical position of the heart is correctly described, and the fact that the heart is filled with blood appears to be well known.[8] The heart is understood as the center of vessels: "In the Heart are the vessels to the whole of the body. . . . Everywhere he feels his Heart because its vessels run to all limbs."[9]

Case 1 in *The Edwin Smith Surgical Papyrus* (c. 1500 BCE) relates the following:

> The beginning of the physician's secret: knowledge of the heart's movement and knowledge of the heart. There are vessels from it to every limb. As to this, when any physician, any Sekhmet-priest or any exorcist applies the hands or his fingers to the head, to the back of the head, to the hands, to the place of the stomach, to the arms or to the feet, then he examines the heart, because all his limbs possess its vessels, that is: it [the heart] speaks out of the vessels of every limb.[10]

Paleopathologists, as well as radiologists who have examined the mummies of Egyptians using computerized tomography (CT), have since found that even in those ancient times, the large vessel (aorta) commonly showed plaques, ulcers, and calcifications, while other arteries were thickened or appeared beaded and were so brittle as to hardly bear handling. If only the ancients had related the *state* of the arterial wall with life and death, what a marvelous pathophysiology they might have recognized![11] The ancient Egyptians may then have grasped, as we do now, that humanity's long reckoning of mortality lies not in the lapse of years but is inscribed upon the walls of our blood vessels. We are as old as our arteries!

two

The Sea Within Us

I have undertaken a journey without maps.

—HERMAN MELVILLE

IN THE SOUTHEASTERN REGION OF ITALY CALLED CALABRIA, WHICH embraces the Gulf of Taranto, the lively port city of Kroton (now Crotone) was on the go.[1] Powerful currents had been running for some time in that part of Magna Grecia between Kroton and the Mediterranean, giving rise to an astonishing and wonderful melting pot. Legend has ascribed Kroton's founding to an athlete named Myscellus who, commanded by the Delphic Oracle, arrived there around 720 BCE and established a colony adjoining the gulf. In time, the city grew into a metropolis of fifty to eighty thousand people, a flourishing nucleus of trade and travel for entrepreneurial men of diverse cults and customs. Pythagoras, the mathematician who gave the world its well-known theorem, settled there and founded his mysterious "brotherhood."

By the fifth century BCE, Kroton was the seat of a medical school and renowned for its doctors such as Democedes, who was personal physician to the Samian tyrant Polycratus and later to Persian King Darius and Queen Atossa at their palace at Susa. The Greek historian Herodotus had this to say about Kroton: "It was by the success of Democedes that the Krotoniats came to be reckoned such good doctors; those of Kroton the best and those of Cyrene the second in all Greece." Cyrene was located on the Mediterranean coast of North Africa, near present-day Shahhat in Libya.

The first natural philosopher to propose that blood moves within the body was Alcmaeon of Kroton, a younger contemporary of Democedes.[2] Homer had described an ebb and flow of the wine-dark sea on Aegean shores, and Alcmaeon imagined the blood's motion to be similar — classicist-author Caroline Alexander points out that the evocative phrase "wine-dark" occurs only a few times in the entire *Iliad*, but it has stuck.[3] Alcmaeon speculated that a leisurely blood flow was set in one direction and then, after an interval, it streamed through the same channels in the opposite direction in endless repetition. With blood moving to and fro in the same vessels, Alcmaeon pioneered a "physiology" of blood flow that would dominate the West for two millennia.[4]

For all its simplicity, the ebb-and-flow model was a positively astonishing idea. Nobody had thought of anything like that before. Still, evolutionary as it was, it was Alcmaeon's *personal* conception. Like many later novel theories, it did not reflect what even the most educated people believed or understood at the time, and it had no immediate effect on how society at large thought about the world. But he was the first to dip into the great riddle of the living body and link the pattern and logic of the external world to the living world within.

The purpose of the blood's motion was to maintain life and consciousness. Alcmaeon drew an analogy between awake and asleep and between alive and dead. He regarded the brain as the center of life and consciousness (another significant original thought for that time) and described sleep as a flow of blood away from the brain, with awakening caused by an opposite motion. Death resulted from a total and permanent drainage. Alcmaeon was also the first to argue that alterations in blood distribution could, under certain circumstances, cause physiological and pathological changes.

All that seemed logical enough because when Alcmaeon was dissecting animals, which he was among the first Greeks to practice, he observed that certain vessels, which we now know as the arteries, were virtually empty. This apparently confirmed that blood was permanently drained after death. At the same time, other vessels (which we call veins) were found engorged with blood. British scholar C. R. S. Harris has remarked that, based on this very slender observation, Alcmaeon may have been the first to distinguish between two kinds of vessels without knowing it, in which case he might as well be called the Father of Western Vascular Anatomy.[5]

What Alcmaeon failed to appreciate was that the relatively bloodless vessels resulted from his savage technique of strangulating the creatures to death. Strangulation constricts the arteries which squeeze out their blood, but not the veins. The smallest arteries of the lungs, called arterioles, actually begin to constrict while the heart is still beating. By the time the animal has passed away, blood has piled backward from the lungs into the right-sided chambers of the heart as well as into all the veins so that they appear engorged at autopsy. In those who die from a collapse of the lungs and a *simultaneous* cessation of the heartbeat, as in drowning, all the vessels remain full of blood.

In fact, many of the differences among later Greek cardiovascular physiologies, whether the researcher believed vessels contained air or blood, or both, can be explained by the method they used to kill the animal they were dissecting: strangulation, exsanguination by bleeding, or drowning.

three

Seeing Is Believing

There is a new visible World discovered.

—ROBERT HOOKE

I N FAIRNESS TO ALCMAEON AND HIS CONTEMPORARIES, ESTABLISH-
ing the "facts" of anatomy through dissection must have been bewil-
dering, given the thick tubes, thin tubes, hollow tubes, solid tubes, and
the different kinds of flesh and fluids that faced the pioneers, the whole
liable to be obscured by blood. Without existing road maps, the first
anatomists could, at most, only record what they saw and make some
guesses about what it all meant.

Nothing about who gave the earliest account in Western medicine
of a *system* of blood vessels is known.[1] The usual suspect is Syennesis,
who lived somewhere in Cyprus sometime before 400 BCE. Nothing
remains of his original text, and he would have remained unknown
had Aristotle not come across a fragment mentioning an observation
ascribed to him. The fragment is also referenced in a treatise titled
De ossium natura (*On the nature of bones*) discovered on the Aegean

11

Island of Cos in the library of the medical school of the Greek Father of Medicine, Hippocrates. It is a compilation of abstracts from ancient authors.

Syennesis described "veins" rising from the eyebrows (!) that proceeded down the body to terminate in the womb or penis. Somewhere along the way, the vessels crossed over to the opposite side of the body. Here is how Aristotle sums up Syennesis's report:

> The thick veins grow as follows. From the eye along the eyebrow, passing down the back, past the lung, under the breast, one from the right to the left, the other from the left to the right. The one from the right goes through the liver to the kidney and testicle, and the one from the left goes to the spleen and the kidney and the testicle, and from there to the penis.[2]

Given that even a cursory glance at the abdomen does not show a crossover of vessels, one must assume that the crossing place, which he called the *chiasma*, must have been in the chest, "under the breast." The crossing vessels seem to fit, as a first approximation, the vena cava and the aorta. The heart was not specifically identified, but later scholars have inferred that Syennesis was most likely referring to that organ as the *chiasma*. Plato famously described the heart as a "knot of veins."

Perhaps the most influential vascular scheme in early Greek medicine comes from Diogenes of Apollonia (c. 435 BCE), who described vessels as conduits distributing a mysterious life-giving substance called "pneuma," later called *spiritus*. Diogenes was raised in the maritime colony of Apollonia Pontica (originally Antheia, present-day Sozopol), which was founded by Greek emigrants who moved southward to Thrace along the western shore of the Black Sea. But he lived most of his life at Athens, where he may have practiced as a physician. He was probably an older contemporary of Hippocrates and exercised considerable influence on some Hippocratic texts. He influenced Aristotle, too, who inserted a précis of Diogenes's account in his own *History of Animals*, which is our principal source.[3] It begins with the phrase "Blood vessels in man are as follows," which is puzzling, since human dissection was not permitted until centuries later. Could he have witnessed a corpse subject to some fearful mishap or a gaping battle wound?

In a passage in the *Iliad*, Homer narrates that "Antilochos rushed Thoön and shore away the entire vein / which runs all the way up the back till it reaches the neck"; Antilochos was the first Greek warrior to kill a Trojan at the onset of that war. In his own scheme, Diogenes, too, described two great vessels that ran at the back on each side of the spine and proceeded under the collarbones up the neck to the brain, and down through the whole body to the legs. From that pair, smaller branches ran to all body structures on their side. He called the great vessel on the left side *splenitis*, and it communicated with the spleen, the left kidney, and the left leg. The opposite one, which he called *hepatitis*, gave branches to the liver, the right kidney, and the right leg. A pair called *spermatitis* ran from the kidneys to the testicles or the womb according to the sex. Diogenes held the blood in the *spermatitis* to be hot and foamlike, reminiscent of the foam in the sea when the goddess Aphrodite was born, and so applied the term "aphrodisia" to the sexual act. The heart had not yet been identified as the center of the vascular system.

Veins were not distinguished from arteries. Both were conduits, called *phlebes*, that transported air as well as blood. A *pair* of them leading to the neck, liver, spleen, kidneys, and genitals were described so that, probably without knowing it, he was another early Greek to record the double system of veins and arteries. The word *arteria* was not used because, in his day, it meant the windpipe. Our word *trachea* for the windpipe is an abbreviation of the Greek *arteria tracheia*. It was much later that *arteria* came to mean the arteries.

ARISTOTLE, AS ALSO HIS MENTOR PLATO, MADE NO DISTINCTION between arteries and veins. But Aristotle did recognize the essential difference in the thickness of the walls of the two great vessels that ran alongside the spine. He differentiated the "great vein" (vena cava) enclosed by what resembled a membrane or skin from the thicker-walled one, the aorta, which he held to be more like a sinew; he was the first to use the term *aorta* in its current sense. He held that pulsations occurred in *all* vessels during life. He was aware that blood was contained only within the vessels when he wrote "there is no blood separately situated by itself, except a little in the heart, but it is all lodged in the veins . . . blood beats or palpitates in the veins of all animals alike all over the body . . . as long as life lasts."[4]

The unequivocal distinction between *all* arteries and veins was made by the natural philosopher Praxagoras (born c. 340 BCE), who lived and practiced at Cos.[5] The place was famous for its wine and silk, and, in later times, a contingent from Cos fought for the Greeks at Troy. Then Asklepios, the Greek physician, arrived from Troy and introduced the art of healing. He was elevated from an especially favored mortal to the status of a god. The island became renowned throughout the Aegean for the medical school established by Hippocrates that flourished between 450 and 350 BCE.

Praxagoras was the first to distinguish arteries and veins as *separate* systems on anatomical grounds based on the relative thickness of their walls, with blood carried in the veins but arteries containing only air (pneuma) and no blood—undoubtedly the most grievous error in early vascular physiology, resulting from the same technique of strangulating animals before dissection that Alcmaeon used. He moved ahead by defining all vessels (and their branches) yoked to the left heart chamber as arteries and those to the right one as veins. He set the vessels even further apart by noting correctly that pulsation, regarded by Aristotle as characteristic of all vessels, was limited to the arteries; he used a word usually translated as "palpitation." He made an advance on Diogenes by recognizing the heart as a distinct organ.

Praxagoras was also among the first to identify the nerves. But he held them to be part of the same network as arteries, which, he taught, were large, hollow conduits when they emerged from the heart. They divided into increasingly narrower vessels until a point was reached when their lumen was completely obliterated. At that stage, they became nerves (*neura*). On the other hand, Aristotle taught that "as the blood vessels advance, they become gradually smaller and smaller, until at last, their tubes are too fine to admit the blood." He did not hold nerves to be the terminal ends of the vessels. Since it was pneuma-containing arteries that became nerves, Praxagoras reckoned that pneuma, now carried within nerves, must cause muscle movement. He held muscle "spasms" and "tremors" to be abnormal conditions manifested by irregularities of the pneuma-containing arterial pulse.

THE FINAL ROUND IN THE GREEK STORY OF VASCULAR ANATOMY WAS played in the cosmopolitan world that Alexander the Great left behind. Talent, and lack of money, led the anatomist Herophilus (c. 290 BCE) to migrate to Alexandria in Egypt from his distant ancestral home at Chalcedon, located on the Asiatic side of the Bosporus (modern Kadikoy, Turkey).[6] Born just two years after the death of Aristotle, he became Praxagoras's disciple.

"He is a wretched pupil who does not surpass his master," Leonardo da Vinci once remarked, a sentiment that Herophilus amply symbolized. He quantified the walls of arteries as being six times thicker, on average, than those of veins, and so established their definitive anatomical distinction.[7] He disagreed with the Praxagorean practice of calling all vessels and their branches arising from the left and right heart chambers as "arteries" and "veins," respectively. He correctly recognized that the large vessel from the right chamber entering the lung was anatomically an artery and not a vein, and so he baptized it the "arterial vein." Adopting the same terminology, the vessels with thin coats, like veins, that went from the left heart chamber to the lung were awkwardly named the "vein-like arteries" by later anatomists, as quoted by the Greek physician Rufus of Ephesus in the first century CE:

Herophilus applies the term "arterial vein" to the very large and thick vessel leading from the heart to the lungs: for in the lungs, conditions are opposite of what they are elsewhere; the veins are there powerful and in nature very similar to arteries, while the arteries are weak and bear a close resemblance to veins.[8]

That distinction and terminology were upheld by his younger Alexandrian colleague Erasistratus, and subsequently adopted by the Roman physician Galen, who used the Latin terms *vena arterialis* or *vena arteriosa* for what we now call the pulmonary artery, and *arteria venalis* for the pulmonary veins. Galen argued *physiologically* for two distinct and parallel vascular systems—he held that the veins, originating in the liver, distributed the nutritive blood made in that organ, and the arteries, emerging from the heart, carried blood *and* pneuma as the "vital spirits."

four

The Life It Brings

As to the breath which enters the nose,
it enters into the heart and the lungs,
and these give [life] to the whole body.

—EBERS PAPYRUS

"AS OUR SOUL BEING AIR, SUSTAINS US, SO PNEUMA AND AIR pervade the whole world."[1] With those words, the "natural philosopher" Anaximenes (c. 550 BCE) enunciated the ubiquity of air (the term *scientist* was not coined until the 1830s). He spelled out a general "world-pneuma" or "world-spirit" that all creatures great and small, including humankind, shared during life, which ensouled itself in the blood as the carrier of life as well as sensation and even consciousness. Pneuma made the act of breathing meaningful because it drew something vital into the body, perhaps the same something that the Supreme Divine Being had breathed into the nostrils of the first man. At death, when breathing ceased, that organism's share of pneuma passed out to rejoin the "world-spirit" from where it was originally drawn. This much

17

we know from the sweetly sad *Meditations* of the Roman philosopher-emperor Marcus Aurelius and others.

Anaximenes, Alcmaeon's predecessor, worked at Miletus in Asia Minor in present-day Turkey and was among the first Greeks to imagine that the external universe was derived from a single, permanent, universal quintessence which he called *aer*. It was the *arché*, the one source, of all animate and inanimate stuff—a formal reductivism of the primal need to find unity in nature.[2] Homer had referred to *aer* as a dark mist or fog; heroes were made invisible by gods in a veil of *aer* just as Trojan prince Paris was during his unequal combat with the Greek king Menelaus. A related word, *aither*, referred to the thin, sunny air and shining sky where divine beings lived. The supreme god Zeus was called "the dweller in the *aither*." Within that brightness, the Olympians moved fast and freely, speeding like lightning through *aither* to float down in *aer* upon the Trojan plain.

Diogenes of Apollonia, Anaximenes's pupil, also argued for material monism.[3] He held air to be the basic cosmic reality because every creature must breathe air to live. Tangible yet intangible, the ubiquitous air permeated everything, including the human body and soul. God was in the air! All other substances were derived from air. In the beginning, Diogenes wrote, echoing Anaximenes, there had been only air, compressible and yet infinitely expansible. The earth was born from the condensation of air into a flat disc and, by further condensation and rarefaction, changed into other things. When air was rarefied, it became fire; when it was condensed, it turned into wind. When it was more condensed, the clouds came to be, when still more, water, then earth, and then stone. Likewise, within the body, air became flesh and the organs. Air, as the primal force, was the agent of thought and the one source of all intelligence and being in the universe and in all living things.

PNEUMA, WHICH SOME SCHOLARS HAVE CALLED "A USEFUL FICTION," had an ancient pedigree.[4] Diogenes held pneuma to be the unchanged atmospheric air that was itself the breath-like carrier of life and self-motion, even of sensation and consciousness. It was significantly represented in Plato's dialogue *Timaeus* and, in later Stoic philosophy, pneuma became an ethereal, almost mystical, concept. It was formalized by the Stoic philosopher Chrysippus of Soli (c. 250 BCE), who considered pneuma to be the vehicle of the *logos*, the divine reason, which was both active as well as organizing. The Roman physician Claudius Galenus (Galen), who lived during the reigns of Marcus Aurelius and Commodus, called pneuma "the soul's first instrument for all the sensations of the animal and for its voluntary motions." If the organs could be regarded as the solid elements of an organism and the "humors" its liquid elements, then pneuma was the airy gaseous element.

The semantic progression from air to breath and from breath to life was already in the minds of thinking men. According to Aristotle, Diogenes was the first Greek to draw attention to the phenomenon of respiration (synonymous, in Greek physiology, to "breathing"). He held pneuma to flow back and forth with blood within the vessels during life. The amount of air that mixed with blood varied: more air lightened blood and led to health and pleasure. In the same way, courage and virtue, and their opposites, were produced ("courage," *corage*, *coraggio*, are derived from the Greek *cor*, which means "heart"). When less air mixed with blood, it became denser and coagulated, resulting in pain and sickness.

Potentially fatal diseases arose from the obstruction of pneuma in the vessels, a concept that prevailed in medicine beyond the Renaissance: pneuma caused illness through "phlegm," which congealed blood by cooling it, or by overheating or boiling it, which also caused a thickening of blood. Phlegm was associated with inflammatory diseases. Giovanni de' Medici, beloved son of Cosimo at Florence, was diagnosed in 1463 of dying of a sudden onset of "phlegmatic complexion" by drinking too much cold water—an excess of water could induce a fatal excess of phlegm! There was, of course, the usual talk of poison, but the lad's imprudence could have furnished sufficient explanation for his sudden demise.

A SIGNIFICANT ADVANCE IN RESPIRATION PHYSIOLOGY TOOK PLACE when the Sicilian Empedocles (c. 450 BCE) speculated that there were pores in the skin. He is the liveliest natural philosopher most roundly fleshed in sources surviving from his time. Born around the year of the Battle of Marathon into a distinguished and wealthy family, he made his home on the southwestern coast of Sicily at Akragas (Roman Agrigentum, Arabic Kerkent), a large Doric colony founded around 580 BCE.

At once a philosopher, prophet, poet, and charlatan (he was denounced as such by the author of the Hippocratic text *Ancient Medicine*), legend has much to do with what is known about Empedocles's persona.[5] As a youth, before he turned to philosophy and medicine, he was actively involved in local politics as an ardent democrat. Stories were told of how, dressed in his usual purple robes with a golden girdle at his waist and a priestly laurel in his long, curly hair, and accompanied by a train of nubile boy attendants, he had awakened a woman from catalepsy who had been "without pulse or breath" for thirty days, and had calmed a maniac through music. His pupil Gorgias, too, witnessed uncanny "magic feats" that can be fairly interpreted as hypnosis. He was not a man of few words; the Latin biographer Diogenes Laërtius (c. third century CE) has recorded Empedocles's opinion of himself:

> When I come to flourishing towns, I am honored by men and women. They follow me in the thousands, asking me where lies the road to profit, some desiring oracles, while others long pierced by grievous pains, ask to hear the word of healing for all kinds of illness.[6]

Certainly, we seem to breathe a different air here! The French philosopher Ernest Renan called him a mixture of Newton and Cagliostro, but most times he was more the adventurer and occultist Cagliostro than Newton.

Still, whatever the uncertainties of his life, there is one feature that continues to command attention—his ill-fated leap to eternity into the crater of Mount Etna, the highest volcano in Europe. Was it an accident or rash intent to prove that he was a god? We shall never know. But the leap into Etna's clouds stirred the Victorian poet Matthew Arnold enough to immortalize that unwise event in modern verse.[7]

EMPEDOCLES'S MOST ORIGINAL CONTRIBUTION TO PHYSIOLOGY WAS his introduction of a double system of breathing: through the lungs via the mouth and nostrils and through his hypothetical skin pores, which he defined as "passages into the outside air, smaller than the particles of the body [blood] but greater than those of air" (Aristotle's words) so that air could pass through, but blood could not. Air entered the mouth and nostrils and was expelled through the skin pores, and vice versa. Breathing was thus a to-and-fro movement of air between the entire body and its surroundings. Then, in an unsurpassed flash of brilliance, he amalgamated that to-and-fro dual respiration with Alcmaeon's Homeric ebb and flow of blood to lay a core framework for Greek "cardio-respiratory physiology."[8] It is the earliest recorded, complete scheme in which movement of blood is linked to the movement of air.

When thin blood rushed toward the "center" (presumably the heart), he wrote, the tips of the skin vessels became empty, leaving room for atmospheric air to fill the void created by the inward blood flow. Air was pushed out when reinvigorated blood returned from the center to the periphery. Thus, blood and breath moved back and forth in the *same* vessels depending on the direction of airflow that thrust the blood with it. The purpose of respiration was to cool the natural heat of the body.

He announced his new physiology in hexameters, again preserved by Aristotle. Bad poetry has not marred its physiological significance!

> *Thus, all breathe in and all breathe out, in all*
> *Are tubes . . . that through the flesh*
> *Stretch surfacewards. . . .*
> *In these have they such openings as give*

Free pass to air but keeps the gore [blood] confined.
When inwards from the surface ebbs the blood,
Tempestuous air with eager wave flows in,
Again, to be expelled with course reversed,
When turns the tide.[9]

It was not clear how blood got to the organs and tissues, whether by attraction from the periphery or impulsion from the "center," or how it returned.

Air and blood moved as one interrelated process. Empedocles did not associate the propulsion of blood with movements of the heart, though he may have speculated that the respiratory inhalation (inspiration) and exhalation (expiration) of air may have been caused by the heartbeat. He held the expansion of the thorax to be simultaneous with the dilation of the heart that, together, sucked in air from the outside to cool the body. The simultaneous fall of the chest and heart, likewise, expelled smoky waste residues by squeezing them out. He gave no description of the heart or the anatomy of blood vessels, let alone the physical association between them. He made no clear distinction between arteries and veins.

EMPEDOCLES SUPPORTED HIS PHYSIOLOGY BY AN INGENIOUS ANALOGY using a common household utensil called a clepsydra, a small, flat-based, hemispherical bronze vessel that curved upward to a long neck with a small round mouth that could be sealed by a finger. Its flat bottom was pierced with holes like a colander. It functioned rather like a pipette. One clasped the vessel by its neck and pushed it down upright into the water, which rushed in to fill the container through the holes. Then close the mouth with your thumb, lift it upright out of the water, and so long as you maintain the seal, the water will not run out of the holes.

The demonstration was memorable, and the explanation and implications obvious. Some pressure in the atmospheric air was keeping the water contained within the vessel despite its holes so long as a seal was maintained. Blood in the body, wrote Empedocles, corresponded to water in the clepsydra. Air was drawn in through the fine skin pores when blood receded from them, as through the holes of the clepsydra, and was driven out into the environment when the blood returned. When blood receded, an invisible but corporeal substance, the pneuma, took its place.[10]

The implications of the clepsydra experiment extended well beyond the confines of cardiopulmonary physiology. It was now shown, experimentally, that all "empty" space was filled and there was no void in nature, and that two substances, water and air, could not be in the same place at the same time. Moreover, matter could exist in a form too fine to be apprehended by sight, and yet, in that form, could occupy space and even exert considerable force. For the first time, the existence of an imperceptible physical universe was revealed, and its significance in Greek science would prove decisive. And from that time on through the millennia, things not seen would become more and more the things of great importance in the scientific world, and more so in our own time.

The imagined Empedoclean skin pores remained the prevailing model for centuries. The comingling of blood and air in the body directly through the pores was, in the medieval world, a microcosm of the relationship between body and soul. For the Elizabethans, the atmospheric air that moved between people could be imbued with spirits of all kinds. Poets sang of air mixed with the beloved's light and spirit that found its way through the skin pores into a lover's vulnerable heart. Theologians worried that demonic forces could gain entrance to people's bodies through their pores. Harvey fought back with reason. "How can the Air so freely, so swiftly pass through the skin, flesh, and habit of the whole body, into the depth, as it can through the skin alone?" he asked. "And how shall Whales, Dolphins, and great Fishes, and all sorts of fishes at the bottom of the Sea, take in the air . . . through such a great mass of water."[11] He vouched for an impenetrable boundary on the skin.

UNLIKE DIOGENES, WHO HELD PNEUMA TO BE THE UNCHANGED AT-
mospheric "world spirit," Plato regarded the heart itself as the "maker"
of a psychic pneuma synthesized from air, which was delivered first
to the brain and then to the rest of the body. It carried vital messages
through blood from the immortal soul in the brain to the limbs and
organs.

Like Empedocles, Plato speculated that the air that entered via
the skin pores replaced air that was breathed out through the nose and
mouth.[12] Air expired through the nose, in turn, pushed fresh ambient
air again through the skin pores into the vessels by a "circular thrust"
(later called *periôsis*, or *antiperistasis* by Aristotle). The ebb and flow
of air thus moved blood back and forth in opposite directions during
inspiration and expiration. In proposing his physiology, Plato chose to
ignore the obvious element of choice during breathing since one can
hold one's breath at will but cannot do so with blood flow as felt at
the arterial pulse. He also ignored the lack of synchronization between
breathing and the arterial pulse.

Having conceived the "circular thrust" of air, he sought a mechanism.[13] The body's heat harbored in blood, he explained, had a natural
tendency to seek its kind in the atmosphere. When warmed pneuma
moved to the exterior through either the skin pores or the nostrils, cold
air from the outside was pushed by a "circular thrust" into the other
entrance "because no void can exist." That which went out was cooled,
that which entered the body was warmed, and the temperature changes
contributed to the circular thrust of air that pushed the blood with it.
"We continue to do this unceasingly, inhaling and exhaling," Plato's
famous pupil Aristotle remarked in his treatise *On Respiration*.

FOR ARISTOTLE, THE PRIME FUNCTION OF PNEUMA AND BREATHING
was the preservation of animal heat. Heat was the *sensu moderno* (motive force) responsible for all physiological processes. Maintenance of
life, and possession of a soul, was dependent on the conservation of a
certain degree of body warmth, stored within the heart. When heat

within the heart increased, the heart itself expanded and the lungs expanded with it like a forge bellows. Cold air was drawn into the lungs with each breath and then passed on to the heart to quench its excess flame, as air "is able to perform the work of cooling, because it also is thin and light by nature, and can quickly penetrate all the parts of the body to cool them."[14]

It is conceivable that he may have imagined the "respiratory" motion to be all one movement whereby the heart expanded from its own rising heat and caused a *simultaneous* expansion of the lungs so that cold air was drawn into both organs together. Breathing was initiated by the *heart*; the heart was the primary *respiratory* organ! The heart, open and porous, would remain the organ of respiration through medieval times to the Elizabethan world. In his play *Antony and Cleopatra*, Shakespeare referred to the heart as a bellows when he made Philo remark that Antony's heart "renegades all temper / And is become the bellows and the fan / To cool a gipsy's lust."[15] It is worth mentioning that we no longer think of the heart as a breathing organ because of William Harvey.

Aristotle was unclear whether the cool air flowed directly into the heart or only into the lungs, which embraced the heart, thus cooling it indirectly. But the function of respiration was now clear enough. The innate heat, which was viewed as an actual flame in the left heart chamber, needed fuel (food), but even more, it needed to be moderated and cooled to keep from burning to exhaustion, and that was accomplished by the cooling effects of respiration.

How air passed between the lung and the heart was unclear. Aristotle viewed the lung as an inflatable sac that contained a complex arrangement of air channels and blood vessels with "hollow" (empty) spaces between them. Following an initial bifurcation, the large windpipe divided into myriads of progressively smaller pipes that eventually terminated in minute branches with *open ends*. Each offshoot was accompanied, perhaps all the way to its terminal open end, by a corresponding offshoot of the vein with its own terminal open end. Maybe air transferred from the terminal open ends of the air pipes into the adjacent open ends of the veins and was then carried to the heart. Air in the lungs, now warmed by its contact with the body heat, was expelled through the mouth and nose. Expired air was always warmer than inspired air. The proof was obvious. Simply breathe out with your open mouth on your hand; the exhaled air is warm!

Praxagoras at Cos made a significant advance over Aristotle. Instead of deriving the motion of the lungs from the movement and heat of the heart, he correctly gave the lungs and thorax their own independent natural abilities to expand and contract. He explained breathing as a complex and continuous process involving four stages, none of which included the heart.

five

Body Flame

*It must be apparent that this element of heat possesses
in itself a vital force that pervades the whole world.*

—Marcus Tullius Cicero

W E NOW PASS ON FROM BREATH TO LIFE.[1] FROM ANCIENT TIMES,
"air" and "fire" were the two elements in which the substance of
life was sought. The Ionian philosopher Heraclitus of Ephesus (c. 520
BCE) was the first to announce a "fire of life" as an innate life-stuff
(*calor innatus*) which resided within the heart. It was the source of vitality and self-motion in living animals and humans. When its flame
was extinguished, the body died, and the lifeless corpse assumed the
temperature of its surroundings.

A living flame within the heart was endorsed at the Hippocratic
school at Cos by the author of the treatise *Peri kardies* (*On the heart*).[2]
Both Plato and Aristotle believed that a living fire was present within the
left heart chamber. It was in the heart that the soul was, as it were, set
aglow: "The Soul subsists in some form of a fiery nature," wrote Aristotle.

To confuse matters, some Hippocratic physicians considered the brain the site of the soul, which led to perennial disputes that continued into the Renaissance. "Although someone does not desire revenge because his blood is burning around the heart," Thomas Aquinas explained, "he is more prone to become angry because of it."[3] The medieval anatomist Mondino de' Luizzi reflected, "As compared to other animals of the same size, [man] has greater heat, the purpose of which is to lift him up to higher levels." He reasoned that the thick walls of the left ventricle preserved the heat within that chamber. Taddeo Alderotti, Mondino's teacher, settled "the heart is the root and the source of the heat." In his 1626 treatise, published just two years before Harvey's work, a Paduan professor of medicine, Pompeio Caimo, fairly represented contemporary thinking when he wrote of "the innate heat . . . as a compound of air and fire, [which] undergoes perennial motion."[4]

IT IS IMPOSSIBLE TO EXPLAIN WHY THE EARLY GREEKS FELT A NEED for a theory of an "innate" heat that was inherent in the living body and indistinguishable from life itself.[5] But it was clear that a generalization as powerful as body heat made possible the unification of otherwise very diverse biological functions. Heat was the cause of self-motion; indeed, of "circular" motion, as well as the motion of the heart. Motion, in turn, generated heat, which was distributed in the arteries. A significant limitation of all early theories of innate heat was the inability to distinguish between heat and temperature, between the quantity and the intensity of heat.

For Diogenes, an indispensable requirement for life-as-action was a specific degree of warm air breathed into an organism. He wrote, "In all living things the Soul is the same thing, namely Air, warmer than that outside in which we are, but colder than that on the sun."[6] Empedocles emphasized an internal body "fire" as the most important quality carried in blood; sleep resulted from a moderate cooling of the blood's heat, the complete extinction of which led to death. His contrarian view that blood, not the heart, was the seat of the innate heat was adopted from folklore: "the blood is the life." He noticed that blood drained from the body became cold and clotted; the perpetual ebb and flow of air and blood, he held, kept the blood fluid within the body.

To the Hippocratic author of *On the Sacred Disease*, pneuma was a derivative of inspired air that first invoked consciousness in the brain before it flowed with blood to the rest of the body and animated it. According to the Hippocratic treatise *On the Heart*, pneuma was the carrier of psychological functions; however, it was no longer derived from the breath but from blood that swelled and vaporized through the flame within the heart chamber. Innate heat continually produced and renewed the processed pneuma, which was the vehicle of the soul. In contrast, breath acquired a cooling function with respect to the heart. Whereas Diogenes assumed that the innate heat was already present in the embryo in a vegetative state and was then raised to the higher forms of soul (sensation and consciousness) after the first postnatal breath, Plato and Aristotle speculated that it was acquired *after* birth by an interaction between ingested food and the innate processed pneuma.

But the nature of the vital heat was not always clear to Aristotle. In at least one section of *De generatione animalium* (*On the generation of animals*), he clearly announced that life was fueled by an unexplained innate heat that was neither a flame nor the elemental heat: "The heat which is in animals is not fire and does not get its origin or principle from fire."[7] The proof was that when death occurred, the cold body retained its constituent elements and the shape of all its parts.

On the other hand, the author of the Hippocratic work *On Ancient Medicine* considered heat to be the least potent power of the body. And later, Praxagoras at Cos, though willing to regard body warmth as an important overall function, denied that warmth was inherent to the body. (He also repudiated an innate pneuma.) Heat, he held, resulted from the conversion of nourishment into blood and was dependent upon the composition of food. Heat was imparted either from the temperature of ingested food or by a certain quality, or essence, within foods that was associated with heat. We are what we eat, he claimed. And we still are! If the amount of heat in the diet did

not suit the nature and constitution of the individual, then indigestion
ensued, leading to sickness.[8]

DIOGENES WAS THE FIRST GREEK TO EXPLORE INEQUALITIES IN BODY
heat, and Aristotle expanded on the Apollonian's findings. The innate
heat, Aristotle held, enacted rules of domination and subordination as
well as general beliefs about shame and honor among humans. The
source of male pride was the body heat that governed the making of a
human. Differences cut most notably across sex, and Aristotle regarded
women as colder versions of men. The passing from cold, weak, and female to hot, strong, and male formed a scale of ascending human worth
that treated males as superior to females despite being made of the same
elements. The fullness, serenity, and honor of naked males as depicted
on the Parthenon friezes made obvious the shame of lesser, female bodies. Hot male bodies were stronger and more reactive than sluggish,
cold, female ones. Greek women were confined to their homes because
the lightless cool interiors were more suited to their physiology than the
open outdoors, exposed to the Mediterranean sun.

The innate heat also determined the mode of reproduction: the
warmer an animal, the more perfect the state in which the young were
generated. Live young were produced by hotter animals; colder ones
produced eggs. Those fetuses well heated in the womb became males;
fetuses lacking in initial heat became females.

As the vital heat presided over birth, so it did over death. Violent
death resulted from a rapid extinction of the innate heat—"the innate
heat has fled him," it was said—whereas natural death occurred from a
gradual exhaustion of heat over time, resulting in an overall body cooling. Everyone could see, Aristotle pointed out, that in the earlier active
stages of life more heat was formed than consumed. Then, as a body
matured, expenditure became excessive so that the stock of heat gradually diminished with age until it was used up and life came to an end.

six

Through the Blood Darkly

And now to proceed to the consideration of the blood.
In sanguineous animals, blood is the most universal
and the most indispensable part.

—ARISTOTLE

SO NOW, AS ARISTOTLE PUT IT, FROM BODY HEAT TO BLOOD.
That blood is essential for life was well known since prehistory, and
for the Greeks it was inseparable from the very idea of life. It had a positive, intrinsic, revitalizing power that was applied in magic and ritual
and in dealings with the gods. When Persephone, the wholly terrible
goddess of death, brought the querulous shadows of the deceased to
Odysseus at Hades, it was not until they had drunk Odysseus's sacrifice
of blood that the souls temporarily recovered life and consciousness.
The Olympian gods, who were immortal, had no blood; *ichor* flowed in
their veins. The same considerations applied to the "vital spirit." It was
held that subtle essences, imperceptible to the senses but vital for maintaining life and body function, were present in blood. That tradition has

continued in our own awareness of vital molecules present in the blood stream that rule and regulate body *homeostasis* (equilibrium). We call them hormones and enzymes.

Air and fire were the two elements within blood that supported life in Greek medical theory. Flesh was formed from blood that was the agent of nutrition, and maternal milk and semen resulted from blood. Contrary to Diogenes who had emphasized air, blood was the primary Empedoclean substrate, and it processed pneuma from the raw and unformed atmosphere. Blood, as the sole carrier of pneuma as well as the innate heat, was the ensouled life-stuff contained in and around the heart, which, itself, was located in the body at the center of the blood flow and in the path of respiration.

In Plato's and Aristotle's physiology, blood was nourishment for the flesh and organs; indeed, from remote antiquity down through medieval times, blood has been compared to wine. That is implicit in Christianity. For Plato, blood was formed from the products of digestion that, on being dissolved, were guided into the vessels. The various substances derived from the plethora of ingested food could be expected to take on all colors when mixed together. Blood was always red, however, because it predominated and was sustained by fire, one of the four basic elements.

The heart was the "well spring" of the blood. "The heart, the knot of the veins and the source or fountain of the blood that races through the limbs, was set in the place of guard," Plato wrote.[1] Significantly, he held it to be the seat of the spirited element of the soul. Thus, the heart was the source of blood not in the context of nutrition or a circulation but rather in its capacity as the seat of the rational part of the soul from which emanated messages carried in blood to the limbs and other body parts. Bloodless parts had no sensation or motion. Plato was unclear whether the vessels were channels conveying messages from the sense organs to the heart, and from the heart through the vessels to the muscles to bring about movement (the nerves had not yet been discovered).

Food, according to Aristotle, was "concocted" in the stomach and turned into a vapor that then rose to the heart, where it was synthesized into blood and distributed in the vessels. Nourishment in blood oozed out to the various body parts and, congealed by the cold, formed the flesh or its equivalent. The earthy constituents in blood, having little wetness, became hard. As their fluid evaporated, they turned into nails, horns, and hooves. Aristotle was acquainted with the two varieties of blood, one florid and pure and of a "cleaner" variety, and the other dark and less clean. Like the innate heat, blood, too, differed in appearance between species and sexes, in different parts of the same organism, and with age, being serous and plentiful in the young and thick, dark, and scanty in the elderly.

A SYSTEM OF VESSELS TO TRANSPORT BLOOD AND PNEUMA WAS DE-scribed in greater or lesser detail in many Hippocratic works. In most instances, it is difficult not to see a deliberate imitation, even exagger-ation, of Diogenes's scheme based around a central *hepatitis-splenitis* pair of vessels.

The Hippocratic treatise *On the Localities in Man* emphasized the movements of blood, including channels of communication between vessels, and also mentioned a "circle":

> The vessels communicate with one another, and the blood flows
> from one into another. I do not know where the commencement is
> to be found, for in a circle you can find neither commencement nor
> end, but from the heart the arteries take their origin, and through
> these vessels the blood is distributed to all the body to which it gives
> warmth and life; they are the sources of human nature and are like
> rivers that [flow] through the body and supply the human body with
> life, the heart and vessels are perpetually moving, and we may com-
> pare the movement of the blood with courses of rivers, returning to
> their sources after a passage through numerous channels.[2]

It is hazardous to conclude, as some scholars have done, that this astonishing description implies a circulation as understood today. The fragment was quoted by the English theologian William Wotton in 1694 in his *Reflections upon Ancient and Modern Learning*. He assessed, fairly, that although Hippocrates "did suppose the Blood to be carried around the body by a constant accustomed Motion . . . he did not know what this constant accustomed Motion was and believed it as a Hypothesis only." Hippocratic physicians were certainly unaware of a literal unidirectional circulation. The fact that many believed that when a vessel was cut blood spurted out from both its cut ends in *opposite* directions completely dispels this idea because such an observation indicates that blood moves in both directions in the *same* vessel, thus precluding a circulation.

ALCMAEON HAD BEEN AMONG THE FIRST TO SUGGEST, ON THE BASIS of his postmortem observation of "empty" arteries, that pneuma might be present within some vessels. Diogenes, Empedocles, Plato, and Aristotle did not comment on specific vessels for the distribution of blood and pneuma; presumably, both were present in all vessels. But there was no doubt that the right heart chamber and the veins always contained blood.

Praxagoras was the first who saw a theoretical need to have two types of vessels to convey the two types of essentials: blood and pneuma. Naturally, he expected the two to have different origins. The venous system, he declared, originated in the liver and contained *only* blood, which was made in that organ from the products of digestion. Veins from the liver then linked up with that right-sided heart chamber that is now called the right auricle or right atrium. On the other hand, arteries arose from the left heart chamber, now called the left ventricle, and were filled only with pneuma, thus attributing a greater significance to pneuma. And in doing so, he echoed a grave error.

In his own physiology at Alexandria, Herophilus correctly held blood to be present in *all* vessels because both arteries and veins had an equal "desire" for nutriment. Whereas veins contained only blood, arteries held a more prominent role because they carried both blood *and* pneuma. That was a vital distinction from Praxagorean physiology. However, because arteries pulsated through an innate property whereas veins did not, he taught that greater absorption by the body tissues occurred from arteries because they were in a better position to push out more nourishment.[3]

On the other hand, his younger Alexandrian contemporary, Erasistratus, adopted the Praxagorean doctrine that only pneuma was transported in the arteries and that all blood was confined to the veins. The left heart chamber was the receptacle for pneuma only, and the right one only for blood. But, unlike Praxagoras, Erasistratus did not make the liver the origin of an independent venous system. Rather, he held the heart to be the origin of two *separate* systems of vessels: the air-containing arteries originated from the left chamber, and the veins carried blood from the right ventricle, presumably through the two venae cavae, with the right auricle considered to be their termination. The right chamber also supplied nourishment to the lung. The liver remained the blood-making organ.

seven

Knot of Veins

The human heart has hidden secrets
In secret kept, in silence sealed.

—Charlotte Brontë

A ND FINALLY, ON TO THE HEART IN GREEK CARDIOVASCULAR
physiology. The earliest description of the heart in Western medical literature is the Hippocratic fragment *Peri kardies* (*On the heart*).
Based on teleological and philological evidence, it is most likely an
addition made centuries later to the Corpus, perhaps just before the
time of the Alexandrian school, although some scholars ascribe an
earlier time. It may have been written by Hippocrates's nephew Poly-
bus. It supports the doctrine of an innate heat in the left heart chamber, where the intellect also resides. Only twelve paragraphs long, the
document deserves study because of the amazing accuracy of some
anatomical observations, including the first mention of the heart
valves. It begins:

[The heart's] shape is like a pyramid, and its color is dark red. It is a very strong muscle, strong not because it is like a sinew or tendon, but because of the thickness of its texture . . . it has two bellies, separate, but contained in a single envelop, one on this side and the other on that. They are not at all like each other.[1]

At first blush, this reads like a modern portrayal. Gray's *Anatomy* starts with "the heart is a hollow muscular organ of a somewhat conical form." But it turns out that for the Hippocratics, muscles were simply a kind of flesh distinguished from other body parts by being firmer and more resistant to dissolution but with no intrinsic property of contraction. Greek physicians spoke rarely, and only in passing, about "muscles" (*myes*); their traditional language was of flesh and sinews. They held muscle to be closely pressed flesh, and that special density equipped the heart to better contain its innate heat. Both the ventricles (bellies), runs the Hippocratic text, have outlet openings guarded by outlet valves whose functions are accurately described as one-way doors. The left ventricle contains no blood but only air, an idea derived from its relative emptiness in strangulated animals. On the other hand, the right heart chamber and *all* the vessels (arteries as well as veins) are filled with blood.

CAREFUL STUDIES ON CHICKEN EGG EMBRYOS SHOWED ARISTOTLE that the first organ to become visible was the heart. It appeared as a tiny pulsating speck on the third day. He saw blood vessels emerging from the organ, which led him to conclude that the heart must be their source. That organ, throbbing visibly, contrasted sharply with the passive, colorless fluid initially contained within it, which could scarcely be looked upon as true blood. Like Plato, he believed the soul resided within the heart.

Convinced that blood appeared first within the heart, Aristotle reasoned that the heart must be where it is created. It was a powerful observation because it contradicted the more widely accepted theories that blood was formed either in the liver or in the brain. The function of the brain, like that of the lungs, was, for Aristotle, a cooling one; it tempered the heat that arose in the heart.

Aristotle conceived the vascular system as a double-trunked tree springing from a common root in the heart and functioning as an irrigation channel. The reason for two large vessels (the aorta and vena cava), he said, was to maintain a bilateral symmetry. Parallel bloodstreams flowed through them out from the heart to all parts of the body. Plato, too, in *Timaeus*, mentioned an "irrigation system" that carried nutriment within the ebb and flow of blood.

Comparing vessels to an irrigation system was illuminating, but it is also what precluded Aristotle and his immediate successors from recognizing the blood's circulation. In an irrigation system, the water, as it runs in its channels, is consumed completely by the crops. Likewise, blood distributed to the tissues as pure nutriment was completely consumed. There could be no return other than waste residues that moved in the opposite direction as an ebb and flow. *There was no need for a circulation.*

ARISTOTLE WAS THE FIRST GREEK TO RECOGNIZE THE HEART CLEARLY and unequivocally as the body's primary organ.[2] Diogenes and Empedocles did not describe the heart at all. Plato, perplexed by the complex anatomy of the heart with the main blood vessels crossing one another, declared the heart "a knot of veins." But Aristotle saw the heart as a waystation for the great vessels as well as a container ("cistern") for blood. The vessels were also containers, but the heart was the first and main receptacle. Instead of bleeding animals to death, Aristotle used strangulation after starvation, which resulted in congestion of only the right-sided (venous) heart chamber. Who could blame him, then, for mistaking the engorged right auricle together with the engorged right ventricle to be a single chamber? He clearly distinguished the auricle separately from the ventricle on the left side.

As a result, he described the heart as having *three* cavities, a serious error that had dire consequences for later physiology. He reiterated that observation in at least three different treatises, but the relationships between the three chambers and the great blood vessels were not consistent, and he put varying amounts of blood in the chambers. "In the heart then," he wrote in *The History of Animals*, "the largest of the three chambers is on the right-hand side and highest up; the least one is on the left-hand side; and the medium one lies in betwixt the other two . . . the aorta is attached to the middle chamber."

The actual number of chambers was not significant. What was crucial was that there be a middle hollow chamber where blood could be contained when first formed. Aristotle's "middle" chamber would have a life of its own.[3] Regarding the heart's function, he was aware that the organ "jumps" throughout life and he knew that the heart was filled with blood. He expressly referred to the difficulty of distinguishing chambers in the hearts of very small animals. He did not conceive the heart as a pump, and he did not mention the heart valves.

The Italian Renaissance anatomist Andreas Vesalius believed that the error in the number of the heart's chambers lay in Aristotle mistakenly "splitting" one chamber into two cavities. Aristotle, he held, had been misled by "the membrane of the left ventricle," that is, the anterior leaflet of the mitral valve, into thinking that the *left* ventricle consisted of two chambers. Galen, on the other hand, held the *right* ventricle to be falsely divided by Aristotle into two separate cavities. Later physicians posited Aristotle's central chamber *within* the partition (septum) that separated the two ventricles, an arrangement that may well have been suggested by Aristotle himself in one of his descriptions of the three-chambered heart.

ANOTHER OF ARISTOTLE'S GREAT ERRORS WAS HIS SOLICITATION OF the innate heat to explain the mechanism of heart function. The Master of the Peripatetics held that the left heart chamber dilated when the fire within it became excessive. Just as overheated milk boils over and overflows, so, too, moist blood, simmering within the "middle" heart chamber because of its internal flame, expanded that chamber and then "boiled over" into the large artery (aorta). It was expansion caused by the foaming of the heated blood that was held to be the principal action of the heart. Hence, "ejection" of blood took place during *expansion* (*diastole*); subsequent contraction (*systole*) was a passive process. We may assume that the boiling blood opened the heart valves, allowing blood to stream out (Aristotle, however, did not mention the valves). That error, whereby blood was ejected during the heart's *dilation*, would be perpetuated for nineteen centuries until rectified by Colombo and Harvey.

According to Aristotle, the arterial pulse resulted from the filling of the arteries from that overflow when the heart was expanded. After such an eruption, the heart was tempered by the cool air in the surrounding lung. Cooling increased the density of residual blood within the heart, which then caused the organ to collapse (contract), and the hot air was breathed out. The force responsible for the motion of blood in the vessels was not described. Whether it was pneuma that formed a motive force to push the blood forward or the blood simply "fell" from the heart into the blood vessel was unclear.

In Erasistratus's physiology at Alexandria, the left heart chamber was, once again, the receptacle for pneuma only and the right heart chamber only for blood. The organ was compared to a brazier's bellows. It was the heart that gave a forward impetus to blood and pneuma; there was no passive Aristotelian "overflow" during dilation using the "boiling milk" analogy. Vital pneuma was thus distributed through the aorta and the arteries. Blood from the right ventricle passed *retrograde* into the right auricle and then into the venae cavae for distribution through the veins. Some blood was carried from the right chamber to the lung through the "arterial vein" (the pulmonary artery).

But how did blood from the right heart chamber enter the right auricle and venae cavae *backward*, given his observation that the heart valves allowed only forward blood flow? No demonstrable solution to this problem was ever given.[4] It would remain for Harvey in the seventeenth century to replace Erasistratean (and Galenic) physiology with a unified theory of blood circulation.

ALCMAEON WAS THE FIRST TO ASSIGN CONSCIOUSNESS, INTELLIGENCE, emotions, and even the soul to the brain and not the heart, an idea that he may have derived from Pythagoras. Diogenes, too, was quite clear that the seat of intelligence and sensation was the brain. On the other hand, the Cnidians seated intelligence in the heart as also the Sicilians, who, like Empedocles, identified heat and intelligence to be in the blood near the heart: "[The heart is] nurtured in the sea of pulsing blood, where especially is what men call thought: for the blood around the heart is thought."[5] Praxagoras concurred, even though he held the heart to be devoid of an "innate" heat.

Both versions gained prominence at the Hippocratic school.[6] There were books in the library at Cos written by physicians who believed in the primacy (arché) of the brain instead of the heart. That concept of a controlling brain created a vertical structure that was familiar to the Greeks, a simple two-part hierarchy of sovereign and subject. Such authors began their descriptions with the vessels in the head. None of them, however, claimed that the vessels had their physical origin in the head in an anatomical sense. One such work was On the Sacred Disease, in which the brain was identified as the organ receiving vessels from all parts of the body. Here, the vascular scheme in the body closely resembled the hepatitis-splenitis of Diogenes. In the Hippocratic treatise On the Nature of Man, the narrative also began with the vessels in the head. The heart was not mentioned; instead, reminiscent of Syennesis, a pair of great vessels beginning at the head crossed the midline in the chest to the opposite side. The possibility of peripheral intercommunications between vessels was raised. All vessels were referred to as "veins" (phlebes), confirming that Hippocratic authors had not distinguished arteries from veins.

INFLUENCED BY HIS SICILIAN FRIEND PHILISTION, PLATO HELD THE heart to be the seat of intelligence and the vital heat as well as the habitat of the spirited element of the soul. For Aristotle, too, the heart remained the center of all the material and mental faculties—the mind, emotions, and all mental processes—a notion that persists in such everyday speech as "heartache" or "heartfelt," "lionhearted" as a synonym for brave, and the phrases "warm hearted," "cold hearted," and "iron hearted." He divided the "mind" into three faculties: perception, memory, and recollection, and placed them all in the heart.

The heart was the cause of movement and even the ultimate determinant of gender! Each of those aspects received more than passing mention, and some of them were quite fully, though confusedly, elaborated. Although it is true that for Aristotle the heart was more important than the brain, the heart was quite dependent upon the brain for practically all body functions, and he concluded that the two organs, together, had supreme controlling power over life. The brain cooled the heart. At the later Alexandrian school, both Herophilus and Erasistratus returned to the brain's sovereignty and emphasized its central role as the organ of sensation and volition. The former took the gigantic stride of demonstrating that nerves originated in the brain, and he differentiated them into motor and sensory nerves.

Thus, by the end of the third century BCE, two notions had evolved for the location of the intelligence and the soul (heart or brain) that would divide physicians well beyond the Reformation. Scholars took part in the debate, as Professor Heather Webb points out, not only by weighing current concepts in physiology but also, and perhaps primarily, by using them as a metaphor for the political state and what it meant, or should mean, to rule a body or body politic. Since antiquity, governance and power in church and state were mirrored in the locus of governance and power in the body, as evidenced by such metaphors as "the crown'd head," "the vigilant eye," and the "counsellor heart."[7] Philosophical and theological formulations of the problem involved rethinking the plurality of the soul and its locations, as well as the belief in the unity of the soul and the single point of its origin. For

Thomas Aquinas, it was all about motion, which determined the soul, and he linked the soul to the only organ, the heart, that showed autonomous motion. His vision of "circulation" referred to the movement of the heart itself rather than the blood or any other entity moving into or out of the heart.

In Renaissance Europe, Giovanni Argenterio, a physician at Turin, perpetuated the belief that many emotional affections arose from the rapid inward and outward movements of heat contained in the heart: "We see that in weaknesses of the mind, fear, dread, anger, joy, sadness, febrile rigors . . . and other universal passions of soul and body," he wrote, "all these occur by the power and movement of this heat [within the heart]."[8] Hannibal Albertini, in 1618, just a decade before Harvey's groundbreaking publication, concurred; during anger, joy, sorrow, anxiety, and worry, "the spirits, running here and there . . . become heated, and when the heat has been communicated to the heart, there occurs palpitation."[9]

"THERE IS A MIND IN YOU NO MAGIC WILL WORK ON," SAID CIRCE THE great enchantress, the Lady of Wild Things, to red-haired Odysseus. That remark hinted at the dawn of that intelligence in the Greek mind that we choose to call scientific. If the developmental stages of science be the piecemeal process by which a variety of familiar notions and observations are added, singly or in combination, to an incrementally growing stockpile, then aspects of that pattern were already in place in the thoughts of the early Greeks. Faced with the task of examining the natural phenomenon of their human existence and of the human body with its air, blood, and innate heat and all the vessels that contained them, they arrived at legitimate possible conclusions that were determined by their prior experiences as well as by the accidents of their investigations.

Inspired by the Homeric ebb and flow of the "wine-dark" sea, they envisioned a life-giving fluid moving within conduits, and systems for the flow of air and blood were set into motion cognizant of a link between pneuma and blood. Arteries were distinguished from veins and a vascular anatomy was defined. The Greeks identified the heart as a vital organ and correctly described the function of the heart valves. Their views on blood, pneuma, and body heat were no less scientific than those that are current today because they were produced by the same sorts of methods, and held for the same sorts of reasons, that now lead to scientific knowledge.

In a way, the Greeks invented organized knowledge and laid the foundations for their inaugural knowledge-based society. There was suddenly a new thing in the world, a systematized intelligence based upon principles that could be reviewed, questioned, and expanded, as witnessed in the Aristotelian Corpus as well as in the systems of air and blood promoted at the Alexandrian school. The Athenians called it *episteme*; posterity calls it "science." And to science they turned for such knowledge of the hazardous future as can be gained by mortals. In Homer's time, it was believed that such information might be sought among the dead. At Circe's instigation, Odysseus reached toward the mist-shrouded land of Hades at the world's edge to divine his own fate. In later times, thinking about the world in the Greek way became the scientific method. And it comes as no wonder that the sciences have rarely existed in the West except among peoples who came under the influence of the Greek mind.

WRINKLES IN BLOOD FLOW

*We must collect the facts and the things
to which the facts happen.*

—ARISTOTLE

eight

A Different Drumbeat

*If you want to get on in life, my boy, lace up
your boots and get out to Alexandria ad Aegyptum
(Alexandria by Egypt).*

—OLD GREEK SAYING

ERASISTRATUS WAS BORN INTO A MEDICAL FAMILY ON THE HO-
meric island of Chios around 280 BCE and was raised at Antioch
in Syria in an environment of court medicine. His mother was the
sister of a physician, and his father, Cleombratus, was physician to
Seleucus I, also known as *Nicator* (the Conqueror). Erasistratus became a pupil of the younger Chrysippus, and he may have studied for
some time, like his older contemporary Herophilus, under the long-
lived Praxagoras at Cos. Others said that he worked at Athens with the
Peripatetic philosopher Theophrastus, who had succeeded Aristotle as
head of the Lyceum.

Erasistratus was among the first to recognize that the function of the
four heart valves was to coordinate blood flow in a single direction. He

also envisioned hypothetical peripheral connections (which he called *anastomoses*) between the very fine terminal ends of the arteries and veins—an anticipation of the capillaries, which were a requirement in his pathophysiology. The *anastomoses*, under natural conditions, were closed. But in case of disease, he held, when there was an excess of blood (plethora) in the veins which then became distended to capacity, the *anastomoses* opened, and blood spilled over *backward* from the veins into the pneuma-filled arteries, resulting in pulse irregularities, fever, and hemorrhage (called the "plethora" theory of disease). It's not clear whether he believed pneuma could overflow from arteries into veins.

In Erasistratus's physiology, the left heart chamber and arteries were receptacles for pneuma only, and the right one and the veins only for blood. His critics were incredulous. How could he even think that only pneuma was present in the arteries! Cut *any* vessel in a living creature, including any artery, and blood spurted out—anyone could see that (clearly, Erasistratus did not enjoy the same academic prestige as Praxagoras, who had successfully made the same argument). Erasistratus explained his way out of that predicament by applying a principle of physics newly spelled out by Strato: a phenomenon called *horror vacui*.

Strato was a materialist. Anticipating Democritus's atomic theory, he held all matter to be composed of minute particles with a tiny void (empty space) between them. He demonstrated that tubes, when immersed upright in a bowl of water and partially evacuated by suction, drew in water. The water particles filled the vacuum in the tube. That principle, referred to as *horror vacui*, stipulated that if a substance was removed from a closed system, another substance must take its place. There could not exist an empty space in nature. As the common phrase goes, "nature abhors a vacuum." Analogies were not uncommon in those days to understand activities in one process in science by seeking resemblances from already explained relations to a similar process in a different branch of knowledge. The physician Erasistratus now borrowed the idea of "following into what is empty" from the physicist Strato. Empty spaces, even in living bodies, must be filled.

When an artery was opened, Erasistratus explained, its pneuma escaped so instantaneously and so completely to the outside that a huge, massed void was immediately created within the entire arterial system. That cavernous vacuum was filled promptly by blood sucked *backward* from the veins into the arteries through their peripheral anastomoses. It was *venous* blood that squirted out of arterial incisions. Blood spurted out of arteries *after* all its contained pneuma had exited and then the whole arterial system was flooded with blood from the veins. Erasistratus also rejected another popular claim that arteries contained blood *and* pneuma—only under abnormal circumstances, such as after arterial incision or during illness, could the arteries contain blood.

ERASISTRATUS USED THE *HORROR VACUI* TO EXPLAIN APPETITE, DIGEStion, and the secretions of bile and urine. Above all, he used it brilliantly to explain *how* blood (or pneuma) moved away from the heart through the network of "irrigating" vessels to the tissues. It was an inspired amalgamation of physics and biology, clearly ahead of its time.

The hypothesis was simple. With use and age, the body's tissues are worn away, which creates empty spaces. Since nature abhors a vacuum, such spaces had to be replaced by something. That was accomplished by the blood or pneuma particles within the vessels. Erasistratus held that the walls of the tiny terminal vessels had extremely fine pores (that he called *kenomata*) through which particles of pneuma or blood could filter out in real time into the most adjacent tissue spaces instantly, and so fill the tissue voids *as they were being created*. Whether that occurred from negative forces (suction) in the tissues created by the vacuum or resulted from successive forward thrusts or pushes of the blood or pneuma itself remained unclear.

In any case, the transfer of blood or pneuma particles into the tissue produced, in its turn, new empty spaces, this time within the vessel itself from where blood or pneuma had just transferred out. That void could only be filled by a forward migration of blood or pneuma particles that were immediately behind those that had filtered out. In that way, as blood or pneuma moved out of the vessels, particle by particle, to replenish worn-out tissue, a forward flow was continuously maintained, particle by particle, within the vessels by the force of myriad vacua that were produced in contiguous segments. Blood flowed, in that way, to the brain, liver, spleen, lungs, spinal marrow, and other organs to renew the substance of each organ. The *same* blood nourished all body constituents.

Here was the most audacious attempt so far to give a unified explanation of biological processes based on mechanical principles. For the first time in the history of biological ideas, a continuous forward motion of blood was not only explicitly recognized but also scientifically explained. *Horror vacui* was likewise invoked to explain the motion of air into the lungs through the nose and mouth and then into the left chamber of the heart. Expansion of the chest created a vacuum that caused atmospheric air to be sucked into the lungs and, in its turn, into the heart through the "vein-like arteries" by a *simultaneous* vacuum created by the heart's own dilation (*diastole*). The "vein-like arteries" (the pulmonary veins) thus carried only air from the lungs *to* the left chamber of the heart.

Logical as that process may have seemed, it was completely erroneous. But backed by the authority of the Alexandrian school, that ingenious Erasistratean physiology driven by the horror of vacuums would hold ground for four centuries.

nine

Prince of Physicians

You imagine that the human mind is looking for truth?
Don't believe it! Man gets on admirably with the false.

—BERNARD LE BOVIER DE FONTENELLE

AELIUS CLAUDIUS GALENUS (GALEN; C. 130–200 CE–C. 220 CE) would be remembered by posterity as the Prince of Physicians.[1] If Hippocrates was a god, then Galen was his prophet. Influenced by his father, Nicon, a wealthy architect, and by the city of his birth, Pergamum, with its famous temple to Asklepios, he received the best possible liberal education in mathematics, logic, language, and philosophy, including all the foremost philosophical systems of his time: Platonic, Aristotelian, Stoic, and Epicurean. As an adult, he continually advocated the proper application of philosophy to medicine. It was rumored that his father placed implicit trust in dreams about his only son's future and, prompted by one of them, added medicine to the lad's curriculum when he turned seventeen. A suitable teacher was at hand. Satyros, an *iatrophilosophos* (physician-philosopher) who was a specialist in anatomy and had authored a learned textbook, had come to Pergamum to reside with the Roman architect Rufinus. The young Galen was entrusted to him for instruction, renouncing pleasures and disregarding worldly matters.

But Pergamum, in the eastern periphery of the Roman Empire, was too provincial for an ambitious youth. The sudden death of his father left him a substantial inheritance. After a dozen years of study abroad, Galen returned home to be physician to the gladiators, a position that offered ample experience in wounds, fractures, and surgery. At the age of thirty-three, after "five seven-month periods" in a flourishing practice, he moved to seek his fortune at Rome, then at the zenith of its power, where he would practice for half a century.

By his own account, Galen gained swift popularity by his brilliance in medicine and his skill at dissection and vivisection. He was appointed physician to the philosopher-emperor Marcus Aurelius and later to Commodus and Septimius Severus, and he won accolades for his brilliant cures of prominent citizens. The Peripatetic philosopher Eudemus became one of his first well-known patients as did the senator and ex-consul Flavius Boethus, a native of Syria Palestina in modern-day Israel. Galen dedicated many books to Boethus, who was clearly the most important influence during his hectic early career in Rome. The satirist Lucian and the geographer Pausanias befriended him, and the wife and son of the philosopher Glaucon became his patients.

"FROM HIS CHILDHOOD, HE HUNGERED FOR ETERNITY." THERE, AFTER all, lay his incontestable claim. But Galen had few of the attributes of greatness. He had a choleric temperament, which he believed he had inherited from his mother. He never married, perhaps for the same reason. Although he set great store on moral virtue, Galen was a stranger to moral fiber and, in many obvious respects, failed to live up to his own ideals. Worse still, he seemed altogether unaware of his failings. He rarely listened to the opinions of others, resolved controversy by simply ignoring it, and was seldom averse to querulous faultfinding. But he was a shrewd judge of men.

Gifted with common sense and an appreciation of what was practical, he knew how to win friends with influential people by unabashedly conceding that he would not have been able to succeed had he not "called on the mighty in the morning and dined with them in the evening." Incapable of accepting competition without degenerating into personal invective, and always grudging of competing scholarship, he was neither loved during his long life nor mourned at death. His scientific legacy bequeathed scores of treatises but not a single immediate pupil. In an age when Rome was like a floodlit stage over which no one with any personality could pass without being recorded in one way or another, much of what is known of his life are his own accounts of himself. Notwithstanding, Galen's medicine *was* monumental, and he knew it. Posterity more than compensated for the lack of recognition accorded to his work by his own professional contemporaries.[2]

FEW HUMANS ARE GIVEN THE OPPORTUNITY TO ALTER THE COURSE of the history of ideas. Of all the ancient Western writers of science except Aristotle, Plato, and the astronomer Ptolemy, no one has had more influence than Galen. Just as the writings of Aristotle and Plato were the basis for science and philosophy, respectively, so Galen's much more voluminous works became the foundation of medicine and physiology. If we omit the Hippocratic Corpus, which was assembled by a school rather than a single individual, Galen's works represent five-sixths of all surviving medical writings from antiquity. For almost fifteen hundred years, he was the main authority on the human body. To be a rational and learned doctor in the time between the fall of Rome and the Victorian era was largely a question of acquiring the knowledge of Galen's works. The proper study of the human body was a study of Galen and not the body itself, and the tyranny of Galen, who enjoyed an almost divine authority, presented the most significant obstacle to progress in medicine. The poet Dante placed him with other virtuous pagans in the safe antechamber to the *Inferno*, and Chaucer mentioned him, alongside Hippocrates, as the model doctor.

The enormous influence of Galen's works, which encompassed all aspects of physiological theory and medical practice, extended well into our nineteenth century, with the teachings of Galenism reigning supreme from Edinburgh to Bombay, from Paris to Baghdad. A highly abridged English version, prepared and published in 1846, remained a standard teaching text. A medical student's oral exam at the University of Würzburg in that same year would have required a commentary on a passage from Galen chosen at random by the examiners. Since then, scholars would, from time to time, advocate a return to Galenic practice. In parts of rural Spain, one may still hear a country doctor respectfully, and affectionately, called *un galeno*. In the Muslim quarter of New Delhi in India, itinerant street healers continue to use the Galenic classification of the pulse in their market practice.

GALEN, NOT UNLIKE ARISTOTLE, WAS AN UNCRITICAL TELEOLOGIST.[3] Nature did nothing in vain. He explained any action that an organ undertook by the end it had in mind (the "final cause").

Every organ, he held, exhibited three linked characteristics: first, a proclivity for a specific action, which led to the next characteristic which was the action itself that, in turn, caused the final characteristic, which was the effect of that action. Every organ also possessed three natural faculties or powers: attractive, assimilative, and expulsive (excretory). What an organ attracted was subjected to an assimilative transformation, and what was not of use or was toxic was expelled. Galen's invention of the natural faculty of attraction was a brilliant one. Just as Erasistratus had done with the more scientific *horror vacui*, Galen now introduced a grand and seemingly unified mythical theory of "attraction" to explain not only the flow of blood but also digestion, nutrition, respiration, and virtually all other aspects of body function and physiology.

He defined the natural faculty of attraction as an affinity between two bodies, or parts of a body, dependent exclusively on a qualitative appropriateness between them, like the affinity between a magnet and iron. Such a natural faculty allowed every organ, or tissue, to attract to itself only those elements in the correct amounts that were most appropriate to its nature. In that way, kidneys attracted urine. That natural attraction superseded gravitational or other physical forces that could be acting to the contrary. Through attraction, every organ could process nutrition from blood to produce substances that were identical to its own constituent tissue, thus allowing its growth and maintenance.

PNEUMA WAS A REQUISITE OF ANIMAL LIFE AND FUNDAMENTAL TO Galen's physiology and medicine.[4] Atmospheric air breathed in was "digested" by the peculiarly frothy and light flesh of the lung. Just as trees drew nourishment through their roots, so, too, the heart "attracted" the processed air from the lung into the left ventricle through the pulmonary veins. Here, the air exerted a cooling effect on the living flame within that chamber and mingled with blood. The resulting admixture, in the presence of the heart's innate heat, was the *pneuma zootikon*, or the "vital spirit." The nature of the interaction between pneuma and blood in the heart chamber was not explained, but vapors and other humors may have been involved. That refined blood was also the appropriate fuel for the heart's contained fire. The vital spirit was distributed in the vivified blood through the aorta and its pulsating arteries to keep an animal alive. All arteries, therefore, carried both blood *and* pneuma. Veins, which sprang from the liver, contained only blood and transported nutritive properties, or the "natural spirit," which was made in that organ. Blood ebbed and flowed in the vessels.

Vital spirit that went to the brain was processed further to form the *pneuma psychikon* (psychic pneuma, or "animal spirit"), which transmitted the higher functions of the soul and conducted sensations and voluntary impulses to the body through the nerves. Galen was unclear whether *pneuma psychikon* and *pneuma zootikon*, that is, the psychic spirit and the vital spirit, were indeed two distinct spirits or were the same pneuma qualified only by their location in the brain and heart, respectively. An additional source of the vital pneuma was the air that could directly enter the arteries through the Empedoclean skin pores. Arteries emerged from the heart and contained blood that *was* the innate heat as well as the vital spirit (*pneuma zootikon*). Both were sustained by the external air during the respiration that took place via the nostrils and throughout the body through the skin pores.

THE PRINCIPAL MANIFESTATION OF LIFE IN HUMANS AND ANIMALS WAS the body warmth. According to Galen, the body's innate heat had its origin in the fetal embryo, which created its own share of heat by using pneuma absorbed through the placenta. The ability to generate heat then continued unabated after birth. The heart was the principal source of the body fire, "the hearth and source of innate heat which vivifies the animal." He saw close structural similarities between the heart and a furnace: the heart had valves, just like a furnace, that guarded the openings that allowed the entrance and exit of blood and air in the heart chambers. The arteries, which conducted heated blood, resembled the pipes and tubes of the central heating system of a Roman home. As in a furnace, the walls of the heart did not participate in the process of combustion.

An experiment demonstrated this combustive process. When the beating heart was exposed in vivo in living creatures, cooling the organ by applying cold packs or irrigating with ice-cold water caused the heart to stop. (Doctors use this insight today during open heart surgery to arrest the beating heart after a patient is connected to a "heart-lung" machine.) On the other hand, application of heat did not cause any obvious damage because the heart continued to beat. It was clear to Galen that cold must have antagonized and neutralized the heart's innate heat, resulting in its arrest.

Heat was the most inflammable of the body's qualities. A little abnormal motion, Galen held, could make it burst into flames within the heart. It seemed quite natural that an understanding of "fire" should lead to an appreciation of the internal body warmth, and Galen found an almost complete identity between a flame and the body heat. Food was the fuel for the body and heat resulted from foods, particularly fatty foods. Food was consumed by the body fire just as oil was by a flame; and just as oil nourished an external flame, so did the fat of cooked meals fuel the internal one. Not only were both phenomena nourished by fuels, both utilized air as well, and got rid of the soot and ashes that were formed as the waste products of combustion. The innate heat needed air lest it be smothered, and it needed air to be cooled lest it burn up all its fuel and injure the body from excessive temperatures. Air, thus, served to cool the heat as well as nourish it. The life process came to an end when the supply of heat and air was suddenly exhausted (as in sudden death) or when toxic fumes eventually extinguished the flame (as in multi-organ body failure).

ten

Coming to Grips

Oft expectation fails, and most oft there
Where most it promises.

—WILLIAM SHAKESPEARE

"IF ANYONE WISHES TO OBSERVE THE WORKS OF NATURE, HE should put his trust not in books on anatomy but in his own eyes." By his own light, Galen was a physician who constantly appealed to experience. From that experience, he cobbled together a very persuasive and completely plausible account of blood flow. The problem was that most of it had no basis in truth.[1]

Galen was fully aware of the four chambers of the heart as they are understood today. However, like Erasistratus, he regarded only the two pulsating ventricles, which he called the *cardia*, as the heart proper. The thin and sinew-like auricles, he held, were the terminal ends of the venae cavae and pulmonary veins and served as reservoirs for blood. When a chest was opened, one could see both ventricles beating autonomously, synchronously, and continually, independent of the brain.

Although a nerve did reach the heart at its base where its membranes enclosed the organ, the pulsative faculty of the heart, Galen observed, had its innate source within its own substance. The heart was the only organ that continued to beat for a considerable time even when removed from the body. In that respect, it differed from all other muscles whose movements were voluntary and controlled by the individual.

Galen refuted Erasistratus's doctrine that arteries and the left ventricle contained only pneuma, and he demonstrated unequivocally by vivisection that *all* vessels, not just the veins, contained blood. When a beating heart, or any artery, was pierced even with the narrowest quill, blood instantly spurted out. Thus, *both* chambers of the heart contained blood in living beings. Arteries visible in the transparent membrane (called the mesentery) in the belly of living animals clearly contained blood. At no time was pneuma seen to bubble out before the exit of blood when segments of arteries were enclosed between ligatures and cut under water.

That observation was significant because it upset the neat Erasistratean scheme that had been the dominant theory for the past three centuries. A drawback to all theories was still the inability to examine body processes in living humans. Unlike Aristotle, who had considered the heart as the *arché*, or primary organ, of the body, Galen proposed a multipolar model. Specifically, he did not give the heart any special position but considered it to be just one of the principal organs along with the brain, liver, and the reproductive glands, each of which had an independent and assigned vital task. Such contradictions between two such trusted authorities as Aristotle and Galen inevitably led, during the Middle Ages and the Renaissance, to the coexistence of two different physiological systems. At the same time, they inspired scholars to offer tremendously creative solutions to reconcile the two accounts.

GALEN ACKNOWLEDGED THAT THERE WERE TWO ANATOMICALLY DIStinct systems of blood vessels. Just as the pulsating heart was designated the *cardia*, so also each pulsating vessel was called an *arteria*.[2] Like

Praxagoras, he defined *any* blood vessel with a pulse to be an artery, regardless of whether it arose from the left or the right heart chamber. The thick-walled pulsating vessel that arose from the *right* chamber of the heart to the lung functioned like the other arteries—he called it the "venous artery" (*vena arteriosa*), now called the pulmonary artery. Likewise, the nonpulsating vessels that linked the *left* heart chamber to the lungs were veins, and he referred to them as "arterial veins" (*arteria venosa*), now called the pulmonary veins.

If there were two wholly separate systems of vessels and all vessels contained blood, then the only plausible explanation must be that they carried two different kinds of blood, which were, indeed, very different in color and texture—one dark, and the other bright red. And they must have different functions. The thick and dark blood found in the right heart chamber and veins, he held, was responsible for nourishing the different parts of the body. The other—thin, pure, and bright—in the left heart chamber and arteries was spiritual and the source of vitality, and distributed pneuma and the vital heat.

Unlike Aristotle, who had so carefully and convincingly established that *all* vessels arose from the heart, Galen declared that each independent vascular scheme with its unique blood must have its own distinctive origin. The veins sprang from the liver, master of its own system and the principal organ for blood formation; the liver was not the "servant" of the heart. Arteries arose from the heart. Blood streamed in veins as well as arteries as a Homeric ebb and flow and was used up by the tissues. Given two separate origins of blood flow, it was *impossible to think of a circulation.*

GALEN DID HAVE A FAIR (BUT ERRONEOUS) IDEA OF HOW THE HEART worked anatomically. Each ventricle had two mouths fitted with valves attached by very strong bonds. When the heart distended, the bonds were stretched and the membranes opened toward the body of the heart itself, allowing blood to pass through. The valves worked as devices to

allow blood flow in one direction. He envisioned a continuous supply of nutritive fresh blood pouring out from the liver for distribution
through the veins. Nothing was conserved, so there *was no need for a
circulation of the blood*. Failure to realize that the true flows in the two
vascular systems were in *opposite* directions—*away* from the heart in
the arteries and *toward* the heart in the veins—kept the pagans and early
Christians from arriving at the truth of a continuous literal circulation
and arguably stunted their scientific progress.

A small portion of nutritive blood went upward from the liver to
the right chamber of the heart and was then passed on to the lungs via
the pulmonary artery. Since its purpose was nutritional, it was used up
by the organ. But Galen maintained that some blood did drip through
the *synanastomoses* between the tips of the arteries and veins in the
lung to negotiate a transit through the lung itself—a bona fide trickle
of pulmonary blood flow.[3] Whether any of that dribble made its way
into the left heart chamber was not described. The right chamber of
the heart, therefore, existed only for the lung's nourishment. After all, it
was reasoned, fish who breathed and fed through gills had no lung and
therefore a right ventricle was unnecessary and was, in fact, missing!

WHEN GALEN DEMONSTRATED THAT THERE WAS BLOOD AND NOT AIR
in the arteries, his challenge was to explain how it got there. If all blood
was formed in the liver and ebbed and flowed only in the veins as a
self-contained system, how could any blood get into the separate system
of *arteries*, especially the vivified blood from the *left* heart chamber?

Galen deduced an ingenious explanation to resolve that dilemma.
The two heart chambers (the ventricles) were separated by a diaphragm-
like partition called the septum. After much study, he noted that in
the right chamber, the two inflow vessels (the venae cavae) appeared
collectively larger than its single outflow vessel that linked to the lungs
(the pulmonary artery) and whose function was solely to carry nutrition to that single organ. Based upon their relative dimensions, it seemed

obvious that more blood was brought into the right heart chamber than left it. Where did that "extra" blood go? Where was it accommodated? On the other hand, quite the reverse was apparent in the left chamber. There, the inflow vessels from the lungs (the pulmonary veins) seemed smaller in their combined dimension than the large single outflow vessel (the aorta) that carried blood to the entire body. Moreover, he believed that the pulmonary veins contained no blood, whereas the pulsatile left heart chamber was full of blood that passed on to the aorta. Where did that blood come from?

The answers were, in a way, self-evident. "Excess" blood from the right chamber had no option but to pass *through the septum* into the left chamber to be vivified. The septum, therefore, *must* have pores that allowed the finer portions of blood to "sweat" through from the right chamber to the left. That was a necessary and pivotal feature of Galen's physiology—*the major portion of blood from the right ventricle must flow into the left ventricle through hypothetical pores in the septum.* That is how venous blood made its way into the arterial system![4] That was Galen's grand synthesis! It was also his biggest blunder.

Here's how he put it:

> In the heart itself, the thinnest portion of the blood is drawn from the right ventricle into the left, owing to there being perforations in the septum between them: these can be seen for a great part [of their length]; they are like a kind of fossae with wide mouths, and they get constantly narrower; it is not possible, however, to observe their extreme terminations, owing both to the smallness of these and to the fact that when the animal is dead all the parts are chilled and shrunken.[5]

After all, as was already known then, blood in the fetus did normally pass in this manner through a visible opening in the septum. However, as we now know, that fetal communication commonly occurs in the septum between the two *auricles*. Because direct observation in living humans was not possible, it was not unreasonable to assume this fetal physiology should prevail in the adult in the ventricles. Galen inferred that this was the only possible way blood from the venous system could pass to the arterial side of the heart and then to the arteries. The septum acted as a filter removing impurities, and the filtered blood became the vital spirit after being processed in the left heart chamber with pneuma and the innate heat.

That was how dark blood from the venous system went to the left heart chamber and became vivified in that chamber to the subtle bright red blood, which also formed the appropriate fuel for the heart's living flame. The interventricular septal pores were a logical necessity despite their inability to be visualized at autopsy. Galen speculated that the pores, open and operational during life, must be obliterated postmortem by the contraction of the flesh of the septum. That was why they couldn't be found in cadavers.

Galen also inferred invisible peripheral connections, called *synanastomoses*, between the terminal ends of the arteries and veins (Erasistratus had called them *anastomoses*). When an animal was bled to death by incising any large artery and then dissected, the *entire* system of *both* veins and arteries was found to be empty, thus confirming that a peripheral communication must exist between them. The common anatomical observation that a vein invariably accompanied an artery also suggested an ultimate communication. But he did not regard those anastomoses to be many, nor did he believe that any significant amount of blood flowed through them during health.

What happened at the invisible terminal ends of the arteries and veins *in the lungs* was unclear. Perhaps the fine terminal ends of the windpipe (*bronchi*) coincided with the fine ends of the veins. Or, maybe, the smallest branches of the bronchi and the veins ended in the tiny interstices of the flesh of the lung and communicated. That suggests Galen believed the bronchioles to communicate directly or indirectly with the termination of the veins. Scholars have proposed that Galen might have imagined the terminals of the veins in the lung to be forked, with one prong linking with the arterial ends and the other joining up with the bronchioles.

Galen asserted one other point. The pulmonary veins (using modern terminology), which connected the lung to the left heart chamber, carried a *dual* gaseous traffic moving in *opposite* directions. Waste, sooty materials in the heart's "furnace," went *from* the left ventricle *to* the lungs for elimination, and cool fresh pneuma *from* the lungs was carried in the same vessels in the opposite direction *to* the left heart chamber to fuel the innate heat. *There was no blood in the pulmonary veins.* Erasistratus, on the other hand, had taught a one-way gaseous flow of only pneuma from the lungs to the heart through the one-way door of the mitral valve. It's notable that both physiologies mistakenly posited no blood in the pulmonary veins.

No unequivocal summary can be given of Galen's pneumatology.[6] As mentioned earlier, he speculated that some blood made by the liver made its way to the right heart chamber and then was filtered through the septal pores to the left heart chamber, where it met the fire in the heart as well as the *pneuma zootikon*. But blood did not form a vehicle for pneuma, nor was pneuma borne in the blood flow. It was drawn through the blood to the peripheral tissues by Galen's invented natural faculty of attraction (*facultas attrahens*).

The *purpose* of pneuma in the arteries was to cool and conserve the innate heat of the body parts, just as the innate heat of the heart was cooled by fresh, processed air from the lungs. Living things were given the breath and a pulse for one and the same reason: the conservation of body heat and therefore of life itself. Pneuma was cold and so prevented the body from overheating. Just as a flame went out in a closed vessel when prevented from contact with the external cold air, so did the innate heat extinguish itself when deprived of the cooling effects of pneuma.

The dynamic expansion and passive falling of the heart was analogous to breathing. But Galen did not hold that organ to be muscular. The rhythmic heartbeat arose through a special faculty of the vital soul, the *vis pulsifica*, contained within the heart chamber, which then spread out from the heart to the arteries. Pulsation of both the heart and the arteries was a *respiratory* movement, the latter involving the Empedoclean skin pores. In his fundamental treatise *De usu partium* (*On the Usefulness of the Parts*), he classified the heart under the umbrella of breathing and compared it to a bellows through which it sucked air into the left heart chamber through the pulmonary veins. The arteries, likewise, sucked in air through the skin pores.

Suction, caused by relaxation of the blood vessels, was the more powerful (negative) force that maintained a forward blood flow. In accordance with Strato's physical law that nature abhors a vacuum, diastolic suction in successive segments of arteries created successive tiny vacuums that drew elements from the immediate proximal segment to fill the local void. In that way, the lighter portions in blood moved forward. Heavier components were drawn by the natural faculty of attraction.

THE ORIGINS OF ALL SCIENCE LIE ULTIMATELY IN MYTH, AND GALEN'S choice of myth was astoundingly successful. Conceptual inventions, such as the heart's septal pores, made the schema work.[7] Most of his ideas were inspired by existing Greek tradition, although rival schools had competed on details. His convincing demonstration that arteries contained blood, not just pneuma, was his most important physiological contribution. His equally dogmatic assertion that the pulmonary veins had no blood was among his most devastating provisions. Still, despite his innumerable dissections and vivisections on animals, including a Barbary ape, and because of his deep study of the physiologies of Hippocrates, Aristotle, and Erasistratus (all of which were speculations), his own system of cardiovascular physiology had come from his imagination. After all, Aristotle, too, had interpreted the structure and function of the heart inaccurately, had asserted that the heart was the center of intelligence, had held that the brain secreted phlegm in order to cool the heart, and had discovered nothing new about breathing.

Endowed with an intellect hardly less comprehensive though certainly less critical than Aristotle's, the immortal Galen brought no new insight to the function of the body. Given his hypothetical septal pores and the mythic natural faculty of attraction, the ebb and flow of blood, his dual origin of blood vessels from two separate organs—heart and liver—with blood being continually generated in the liver and used up at the periphery, compounded by Erasistratus's incorrect inference that the pulmonary veins carried only air from the lungs to the heart and Galen's equally erroneous description of a bidirectional gaseous traffic of air and sooty wastes in those veins—all those concepts, taken together, stood as formidable obstacles to preclude an earlier discovery of a true circulation. His was a pretty theory; regrettably, it raised more difficulties than it solved, and it contradicted more facts than explained them. The riddle would be unread for over fifteen centuries.

NEW HEARTS FOR OLD

No one can produce a theory so sound but that facts, time, or use may not bring forth something new to show that one's fancied knowledge to be ignorance, and that one's first judgement is repudiated by experience.

—Publius Terentius Afer (Terence)

eleven

Brave New World

They must find it difficult, those who have taken authority as truth, rather than truth as the authority.

— GERALD MASSEY

O F THE WHOLE STORY OF CREATION, MICHELANGELO PAINTED only the creation of the first Man. For him it was not, as in the story itself, the last and crowning act of a series of developments but the first and unique act: an epigrammatic statement of the divine potential of man. Fair as the young men of the Elgin Marbles, the Adam of the Sistine Chapel is unlike them in his self-contained balance and completeness, with languid beauty wrought from within upon the flesh. Here was posterity's greatest inheritance from the Renaissance: the confident assertion of man's everlasting destiny, though impeded by the Fall, to become the child, companion, and embodiment of God.

The movement that began the new era was twofold: partly a rediscovery of antiquity, and partly the coming of a "modern spirit," with its realism and its appeal to experience and nature. The painter Raphael represented the return to antiquity, and the polymath Leonardo da Vinci, the modern spirit, with a return to nature with its perpetual surprises, finesse, and delicacy of operation. Together, they gave birth to a transformation of life itself, etched and molded after the thoughts and experiences of a world gone by, replete with the animalism of Greece, the lust of blood-bolstered Rome, the reverie of the Middle Ages, and now, the sins of the Borgias.

LEONARDO WAS PERHAPS THE ENCYCLOPEDIC SCIENTIST WHO MOST embodied the spirit of his age.[1] In those days, the rationality of physiology and medicine was supported by its anatomical basis, and his relevance to our story rests upon the anatomy and pathology of the human and bovine hearts and vascular systems that he described in his notebooks. His illustrations were far superior to any before or since during the Renaissance and would have laid a superior foundation for Vesalius and Harvey had they been available.

Leonardo's origin and misfortunes alike stemmed from what his world condemned as a *mésalliance* between his father, a Florentine notary and landlord, and a teenage peasant girl named Katrina. An illegitimate child raised without maternal love and renounced by his stepbrothers, accused by the signory of homosexuality, a man without a father or fatherland, he remained forsaken and misunderstood. No portrait of his youth remains, but all evidence makes us believe a keen and puissant nature with a character strong enough to balance the disadvantage of his birth. For a time he labored manfully to earn his living, but nature pleased him more than humankind. From the first, he was an artist and painter, a draftsman and an engineer. He was fascinated by the appearances of things, impressed by their contours, surfaces, and substances. Perhaps the scientist grew out of the artist: alongside his drawings, sometimes on the same page, were notes in which this myriad-minded man puzzled over the laws of kinetics and the operations of nature.

He was the fullest man of his time, and he tried his hand at almost every science. His study of the heart sprang first from his wish to represent the surface anatomy of the body, to uncover and understand, layer by layer, the muscle under the skin and the bone beneath the muscle. Then his curiosity passed below the visible surfaces to mechanisms. No single motive unified his energies, which pursued a confederacy of interests or dispersed themselves through different provinces of investigation.

He took enthusiastically to mathematics, was fascinated by astronomy, wrote a hundred pages on motion and hundreds more on heat, weights, acoustics, color, hydraulics, and magnetism. Armed with the great text of Theophrastus, he turned to natural history and devised the best botanical classification before Linnaeus. He inaugurated the field of comparative anatomy by studying the limbs of man and beast in juxtaposition. No contemporary knew more of the strengths of materials and the modes of building; no one studied as profoundly the power and currents of air and water, the wings of birds, and the prancing of horses. As a young boy in the street market at Florence, he had bought caged birds just to set them free and study their flight. He helped create the science of chemistry, understood the general nature and implications of fossils, designed inventions as varied as the rifled cannon, steamboats, submarines, and flying machines, and marveled at the human eye and the phenomena of light and sight. From the theory of shadows, he derived the *Virgin of the Rocks*; from an analysis of light, the mystery of the *Mona Lisa*; from his treatise on movement, the frenzy of the *Cavalry Battle of Anghiari*; and from the casuistry of the passions, the drama of *The Last Supper*.

Later writers saw in those efforts an anticipation of modern mechanics; to him they were rather dreams thrown off by an overwrought and laboring brain. Two ideas, indelibly fixed within him, had touched his mind since childhood beyond the measure of other impressions — the smiling of women and the light and shadow that contributed to their visible expressions, and the motion of great waters. All his life he followed the solemn effect of moving water, springing from a distant source among the rocks in the *Madonna of the Carnation*, passing as a little fall into the treacherous calm in the *Madonna of the Lake*, stealing out in a network in the *Mona Lisa*.

Leonardo's generalizations proceeded through analogies. Man was a microcosm as the ancients had said, a smaller world, *Mondo minore*. The human body was composed of the four elements, as was the world, and had bones just as the earth had rocks, and a pool of blood in which the lungs swelled and collapsed just as the ocean tide rose and fell. And as the ocean filled the body of the earth with infinite veins of water, so, too, the veins of the body arose from a pool and went branching to all the members. Only, the stable earth lacked sinews that were made for movement. Just as life and spirit made the body a living organism, so it was with the world where nothing was born save where there was life.

LEONARDO DESCRIBED THE ANATOMY OF THE HEART IN 1513, NOT only in words but also in drawings that excelled beyond anything hitherto done in that field.[2] He was the first to represent the organs and limbs in cross section (like a modern CT scan); the left and the right were commonly reversed in his notes as a result of his habit of mirror writing.

Drawings and descriptions in his notebooks confirm that Leonardo studied the heart and blood vessels from the dissections of animals, principally oxen, although, by his own account, he had dissected thirty corpses in his lifetime. He stressed the heart's central position in the body, midway between the brain and the testicles, and commented that its obliquity to the left, when added to the spleen, balanced the liver on the right. In other drawings he outlined the relation of the heart chambers to the great vessels. The superior and inferior venae cavae were carefully shown to separately enter the right auricle, which he referred to as the "gibbous part of the heart." The lung's *vena arterialis* and *arteria venalis* were correctly identified, and the aorta (*arteria aorto*) was shown emerging from the left ventricle. He drew the outer surfaces of the heart and, by studying the organ from four views (front, side, back, and cross section), demonstrated its cavities, their form and projections, and the venous and arterial openings with their respective valves. In one of those, the heart was shown with its coronary arteries coursing over its surface.

Leonardo initially believed in Galen's theory of the origin of the veins from the liver and even made a drawing of the venous system as a whole, titled "The Tree of the Veins," in which he sketched a man with colored veins arising from the liver. He eventually dismissed that concept on the basis of his own observations, which he found corroborated by the writings of the Persian Avicenna and the Italian Mondino's medieval text. In several notes, he jotted that both arteries and veins arose from the heart, which he called the root of all vessels, since the maximum thickness of the vessels was at their junction with that organ. He regarded both arteries and veins as conduits for the distribution of nourishment and heat but devoted considerably less attention to the arteries in deference to the greater importance attached to the veins by Galenic authority.

The vessels in the neck he called the apoplectic vessels (the term *carotid* means "to stupefy" in Greek) because, he recorded, "if you compress the 4 vessels of either side which are in the throat, he who has been compressed will suddenly fall to the ground asleep as though dead and will never wake of himself; and if he is left in this condition for the hundredth part of an hour, he will never wake, neither of himself nor with the aid of others."[3]

Unlike Galen and the Greeks before him and Vesalius and the Italian anatomists after him who all considered the heart as two-chambered, Leonardo correctly divided the heart into four chambers. The two upper he called the "external" ventricles or "additions" (*additamenti*, a medieval term used by Mondino), and below them were the two true "internal" ventricles. Galen and the Arabs had uniformly regarded the auricles as lying outside the heart (hence, "external") as dilated terminations of the great veins (the venae cavae). The auricles, Leonardo held, were designed by nature as expanding pockets to buffer the percussion made by blood when it was forcibly driven out of the ventricles, similar in function to the bundles of wool placed on the bulwarks of warships to soften the impact of enemy cannon. He sketched the septum between the chambers from the side and in transverse section as a solid structure. He admitted that he had been unable to find the septal pores (*meati*) as described by Galen but did not deny their functional presence because of his great respect for authority. But in one drawing, he did indicate a patent congenital opening in the auricular septum (now called the foramen ovale). "I have found . . . a perforating channel," he commented, "which I note here to see whether this occurs in other auricles of other hearts."

He confirmed the arrangement of longitudinal, transverse, and oblique fibers in the main mass of heart muscle, and further recognized an inner set of intrinsic muscles—the *trabéculae* lining the ventricles and the pectinate muscles lining the auricles. He noted that the papillary muscles of the heart (*musculi papillares*) divided into two parts, with each attached to the sinew-like *chordae tendinea* that fastened to the cusps of the valves and held the heart valves in place. He made beautiful drawings of the papillary muscles and chordae that were fixed to the bicuspid and tricuspid valves between the auricles and the ventricles. In some of his diagrams, he outlined a muscular band that he called the *catena* (now known as the "moderator band") that originated in the septum and attached to the base of a papillary muscle in the right-sided chamber. In his opinion, it served to prevent the heart from dilating more than necessary.

The papillary muscles, he held, also served to prevent excessive dilation of the heart. But he failed to recognize that they were indeed integral to the main mass of the heart muscle, and they all contracted and relaxed in unison. Instead, he proposed that when the heart was dilating, there was an independent active *contraction* of the papillary muscles that pulled the valves back and opened the doors between the auricles and the ventricles. The force of blood in the ventricular chambers during contraction then pushed the valve cusps back together, shutting them. When the auricles contracted, blood was driven through the valve openings into the ventricles, which were simultaneously expanded.

OF ALL THE BODILY MOVEMENTS, LEONARDO FOUND THE HEART AND its valves the most perplexing. His drawings, experiments, and deliberations reflect his unceasing effort to reach the ultimate truth regarding that abstruse problem, and he expended much of his time as well as his knowledge of the laws of physics in their observation. He recognized that the heart was a principal and powerful muscle but preferred to define it as a "vessel" made of thick muscle, vivified and nourished by its own artery and vein. The heart motion, spontaneous and independent of nerve function, originated in the heart muscle itself, and it occurred along the heart's longitudinal axis. At one time, he even proposed the source of that impulse to be in a papillary muscle of the left ventricle.

He studied the motion of living hearts in dying pigs by thrusting into that organ an instrument called a *spillo*, which was used in Tuscany to draw wine from casks. When the heart moved, the instrument moved too, with its handle going downward when the heart relaxed and dilated, and the reverse motion taking place when the heart shortened. The conclusion was revolutionary. For the first time, he confirmed that the heart shortened during contraction and lengthened during relaxation. The entire weight of tradition had advocated the contrary. Unfortunately, Leonardo failed to grasp the enormous significance of his observation. When the heart ceased to move, he observed that the handle of the *spillo* became stationary in the exact middle of the two extremes, thus suggesting (erroneously) that the heart, at death, ceased to beat in a position midway between contraction and relaxation.

He studied the valves of the heart after the large vessels had been filled with wax, thus allowing their true form to be appreciated. The

heart's function was analyzed as mere muscle action by observing the
organ with its valves cut off, and he concluded that closure of the heart
openings proceeded more quickly and completely in the presence
of the valves *and* their papillary muscles than by action of the heart's
contraction alone. He constructed a hydraulic model to observe the
valve cusps when fluid was passed through them, and he accepted some
degree of leakage at the valves to be normal.

LEONARDO WAS ONE OF THE FIRST TO EXPRESS NATURE AND BIOLOGY
in terms of the mathematics of the Greeks, especially Archimedes,
which was being newly rediscovered. To understand blood flow through
the aortic *ostium*, he experimented with models. Hydraulics, or what
Leonardo called "the matter of water," was a subject of abiding interest,
and he extended his mathematical reflections to the problems of blood
flow. He analyzed the three-cusped aortic and pulmonary valves in
terms of the properties of a triangle, and the root of the aorta above the
aortic valve in terms of "the concavity of the semicircle [that] reflects
the percussion of the blood with a large and speedy momentum towards
the center of the triangle a b c where it presses against the apex of the
[valve] cusp," he wrote, explaining an adjacent diagram.

He constructed a wax cast of the ventricles and their vessels over
which he made a gypsum cast, and from that a glass one through which
he could observe the vortices made by blood when contracting ventricles drove it into the aorta and pulmonary arteries. He recorded his
findings in drawings using the example of a spouting pipe. When blood
entered the aorta, it struck the wall of that pipe and divided at its top-
most edge into an ascending and descending stream. The descending
part made a spiral curve that traveled down again to the root of the
aorta onto the surface of the semilunar aortic valve, thus stretching the
valve and closing it. The same wave then turned upward in a retrograde
movement and ended in a reflux (reflected) vortex.

Leonardo concluded that the speed of blood entering the aorta varied proportionately to the caliber of the vessel, similar to the dynamics

of water running in a pipe. The pulsation of blood ejected by the heart created a wave that went through every artery and could be heard by placing the ears at the temples. The aortic semilunar valves opened at the ingress of blood into the aorta and closed with its withdrawal.

LEONARDO WAS THE FIRST TO APPRECIATE THE RELATION BETWEEN heat and friction, and he applied his new discovery to the heart's action to get around Galen's mysterious "innate heat." Flux and reflux created friction that, in turn, generated heat. He wrote:

> The heat is generated through motion of the heart and this is shown
> because the more rapidly the heart moves the more the heat is in-
> creased, as the pulse of the febrile [patient] teaches us. . . . And so,
> by such flux and reflux made with great rapidity the blood is heated
> and subtilized and becomes so hot that but for the help of the bel-
> lows called the lungs which by dilating draw in fresh air . . . and
> cools them.[4]

Blood was driven by a whirling movement in the ventricular chambers, and as it moved against the walls, the resulting friction heated it. He jotted in his notebook, "Heat is produced by the move-ment of the heart, and these manifests itself because in proportion as the heart moves more swiftly the heat increases more."[5] The genera-tion of heat, and not the mechanical transport of blood, was, for him, the essence of the heart's action. The faster the heartbeat, the greater the friction from the churning blood and the greater the heat produc-tion. That explained the well-known association between a fever and a rapid heart rate.

When the ventricles contracted, some blood leaked back into the auricles *before* the valves closed entirely. That observation was funda-mental to Leonardo; indeed, he asserted, the flux and reflux of blood between the upper and lower chambers also contributed to heat the blood and make it more effervescent ("subtil") in quality. Heat, in turn, gave life to all things just as the warmth of a hen gave life and birth to her chickens and the sun imparted life and blossoming to flora.

LEONARDO WAS AWARE OF THE THICKENING AND TORTUOSITY OF blood vessels in the elderly. He observed the hardening and narrowing of arteries from the dissection of *il Vecchio*, an old man who "a few hours before his death, told me that he had passed one hundred years, and that he was conscious of no failure of body, except feebleness. And thus, sitting upon a bed in the hospital of Santa Maria Nuova at Florence, without any untoward movement or sign, he passed from this life." Leonardo performed an autopsy on the man who had passed away quietly before him "to see the cause of a death so sweet."

It was from *il Vecchio*'s postmortem that he reported so vividly the condition now termed "arteriosclerosis." Most striking was his description of the calcification of arteries: "I have found stones in the vessels which pass beneath the clavicles of the chest. These were as large as chestnuts, of the color and shape of truffles." He also noted that the thickening of the vessel walls was, at places, sufficient to obstruct blood flow: "[The walls] acquire so thick a covering that it contracts the passage of the blood," he wrote. Coronary occlusion was mentioned, perhaps for the first time in the history of pathology, and Leonardo concluded that *il Vecchio*'s death had occurred "through lack of blood and deficiency [in] the artery which nourishes the heart."

In his physiology, Leonardo was Galenic. He endorsed an ebb-and-flow motion of blood and believed that some blood that ebbed and flowed in the right ventricle passed through the interventricular septum (which he called the "sieve" of the heart) to become refined in the left ventricle to form the vital spirits, giving rise to the natural heat of the body. Like Galen, he also believed that the vital spirit was carried from the left ventricle to the brain, where it was transformed into the animal spirit.

But his was a tale of labor lost. When Leonardo sank to rest at the Chateau de Clou under the protection of Francis I after nineteen years of wandering, he left less than a score of paintings and not a single complete statue, machine, or book. But his observations and experiments that filled thirteen volumes of manuscript and all his notebooks, five thousand pages of writings and illustrations, would have little influence on the evolution of any scientific field for over two centuries. The plethora of drawings and notes were hidden and unread, consigned first to the Ambrosian Library at Milan and later to the Royal Library at Windsor, finally to be uncovered as a testament of the greatest mind of his time.

twelve

The Proper Study of Man

*Hence, instead of Man Thinking, we have the
Bookworm. Hence, the book-learned class,
who value books as such; not as related to
nature and the human constitution.*

—RALPH WALDO EMERSON

THE YEAR 1543 NEED NOT HAVE BEEN REMARKABLE, AND FOR
many people around the world it meant little. But for Christendom
it meant a great deal. For the year 1543 saw the appearance of the astronomer Copernicus's *De revolutionibus,* and the printing presses of
Jay Oporinus at Basel gave to the world a folio volume *De humani corporis fabrica libri septem* (*The seven books on man's bodily works*) written
by a dynamic young superstar named Andreas Vesalius: an anatomist by
profession, a humanist by passion.

Perhaps it was not entirely coincidental that those two great works
appeared only a week apart. Descended from a family of physicians
from the town of Wesel by the Rhine, Andreas van Wesel (Vesalius) was

an ambitious twenty-nine-year-old anatomy tutor at the time of its publication, his material freshly taken from the dissecting tables. Mikolaj
Kopernik (Copernicus), on the other hand, was a dying recluse who
had lived shrouded in secretive and uncertain twilight as canon of the
cathedral at Frauenburg (now Frombork), Poland, untouched by the
joys and sorrows of the world. Thanks to nepotism (his uncle, a bishop,
had helped him secure his tenured position), he had made for himself
an existence full of peace and happy tranquility. Among the Latin mot-
toes that adorned his quiet den there was one in which he had heard
the voice of a life uncorrupted, a voice of pure lyricism without a trace
of falsehood: *Domi manere convenit felicibus*, "the happy do best to
stay at home." After three decades of laborious computations in that
muffled solitude, he was hesitant to the last to enunciate his great hypothesis—that the earth was a "wanderer" while "the sun, as if sitting
on a royal throne, governs the family of stars which move around it."
Its immediate impact was significantly reduced because a fearful editor had replaced Copernicus's introduction with a milder, misleading
preface of his own.

Both Vesalius and Copernicus were voyagers. Both men forfeited
the orthodox hopes that had sustained the Christian worldview through
past centuries. Neither's work was "revolutionary" in content; their
achievements proposed important changes indeed, but they were not
the shattering blows they are commonly taken to be. Vesalius was no
more successful in escaping the limitation of Galen's physiology than
Copernicus was in departing from the formal astronomical system of
perfect celestial circles. Each, however, in his own way, innovated a
change more in scientific attitude than scientific invention and profoundly influenced the generations that followed, inspiring trains
of activity that, in their turn, led to evolutionary advances. As things
played out, the year 1543 would come to mark the beginning of that
period of transition and contrasts in science now baptized the scientific
"revolution."

WITH RARE HISTORICAL NEATNESS, ANDREAS VESALIUS WAS BORN IN Brussels as the last night of 1514 was passing into the first morning of 1515.[1] The boy's life overlapped with Erasmus of Rotterdam, who had "laid the egg of the Reformation," with Luther, who hatched it, with Henry VIII, who manipulated it for personal gain, and with Charles V, who labored hard to exterminate it.

Vesalius hailed from a noble family, which he gave up to win eternity. His known paternal lineage began with his great-great-grandfather, Peter van Wesele, and three generations of male ancestors had been physicians to royalty—to the emperors Frederick III, Charles the Bold, Maximilian, Philip the Fair, and now Charles V. But his grandfather had never married his grandmother, so his own father, born out of wedlock, was reduced to the position of royal apothecary at the court of Margaret of Austria. While yet a boy at Bovendael in Belgium, which was near a wooded area where criminals were hanged, Vesalius taught himself dissection upon the bodies of mice and men, passing his night hours in the company of corpses quartered and flayed and most horrible to behold. Indeed, no creature could be safe from him!

He studied first in his native Flanders at Louvain, then known as the Belgian Athens. He took up residence at the Castle School, the new humanist college, where he was grounded in Aristotelian orthodoxy and natural philosophy. Among its famous residents were the Italian philosopher-physician Pietro d'Abano as well as the premiere humanist of that age, Desiderius Erasmus. He then transferred to Paris at the age of nineteen and rounded up his studies at Padua.

PARIS IN THOSE DAYS WAS A BASTION OF RELIGIOUS ORTHODOXY WHERE Luther's doctrines were thoroughly condemned in 1521. The French reformer Louis de Berquin was burned alive for heresy as recently as 1529. The Catholic syndic of the Sorbonne, Noel Beda, even regarded learning Greek a heretical activity. By Vesalius's time, humanistic scholarship was already making headway in France. Francis I was on the throne, and, at the royal court, the new Renaissance spirit was in full flower. Leonardo da Vinci, Benvenuto Cellini, and Andrea del Sarto, among others, regarded him as their patron. But unlike the French court, the French university, with its medieval colleges and tradition, was slow to welcome change.

The study of medicine had been introduced at the schools in Paris by the close of the twelfth century. The first official record of medical teaching dated to 1213 when Pope Innocent III gave the chancellor the right to license lectures on *physica*. The hospital of the Hotel-Dieu was enlarged and renovated. By the sixteenth century, Paris had become the center of a renewed interest in Galen and a stronghold of conservative Galenism. There had been little previous activity in anatomy in Paris, and what took place now was almost purely an outgrowth of the newly available Galenic treatises, as well as those viewed through the lens of medieval Aristotelian scholasticism and Arab authorities. In 1526, the Faculty of Medicine purchased the complete Greek edition of Galen published by the prestigious Aldine Press of Venice. By 1528, Paris printers were producing Galenic texts and translations at the rate of two editions per month! The Parisian Galenists not only busied themselves with translations of Galen's anatomical writings but also sought to make them the subject for a teaching program based upon dissection.

When young Vesalius entered those portals in the summer of 1533, he was swiftly caught up in the activity to master Galen's writings and learn the dissection techniques. He counted among his teachers several distinguished conformist physicians, including Johann Guinther of Andernach, regent doctor and primary instructor in anatomy at the university and among the first to translate Galen's anatomical treatise from Greek to Latin. Others were the physician-philosopher Jean Fernel, who was the primary professor-in-ordinary, and Jacques Dubois (latinized to Jacobus Sylvius), a dour and foul-tempered bigot who began his medical studies while in his fifties and gave his name to many body structures that they still bear. Although Sylvius was crude and cheap, Vesalius credited him as a genuinely skilled anatomist.

Sylvius, a native of Amiens, lectured by the book. He was so ardent a follower of Galen that he held any structure in the human body that was different from Galen's description to be a degeneration. He encouraged autopsy in those who died from disease so that, by recognizing the cause of the malady, others may be better healed. Notwithstanding, actual dissection and public anatomies were rare in that bastion of book learning, and it is unlikely that Vesalius attended, or assisted at, more than three public anatomies.

WHEN HE BEGAN STUDY AT PADUA, VESALIUS WAS STEEPED IN THE learning of the now easily available Galenic texts but had slight experience of the human body. He was unacquainted with the vast technical literature on anatomy and knew little about contemporary Italian anatomists. It was not until 1540 that he realized that many of Galen's descriptions had been taken from animals. Alongside the sanctioned dissections, Vesalius investigated the human body on his own. From gibbets, cemeteries, and malodorous charnel houses he assembled and carefully studied the bones and organs of the sinful dead, and the burial grounds attached to the nearby Church of the Innocents provided exceptional riches to the inquisitive youth through the corpses of criminals and worn-out paupers as well as bodies wasted by disease. "I climbed the stake and pulled off the femur from the hip bone," he wrote. "While tugging at the specimens, the scapulae together with the arms and hands also followed. . . . I carried them some distance away and concealed them . . . until I was able to fetch them home bit by bit."[2] That inner preparation that marked his youth, as well as his near silence about it, bore witness to his lifelong single-mindedness.

One might have called Vesalius the Luther of Anatomy because he made it a point in his anatomy to disagree with the ancients. The exciting events of Luther's spiritual rebellion at Rome had happened during Vesalius's youth while he was in the Netherlands and, inspired by it, his vigorous, teeming mind began a parallel advance in anatomy as he taught his students to see a different human body. But the substantive content of what he conveyed was still Galenic. The connection between Vesalian anatomizing and Lutheran Protestantism became striking when Melanchthon, known as Luther's Bulldog, was elected preceptor of Germany and made anatomy totally God-centered through his reforms at Wittenberg. Anatomy, he held, led to the knowledge of God. In 1552, Melanchthon composed a poem at Nuremburg in praise of the human body and inscribed those verses on the title page of his personal, closely read, and annotated copy of *De fabrica*, Vesalius's own great book on human anatomy.

AFTER BEATING THE UPPER AIR FOR A WHILE, THE CHARISMATIC VESA-lius broke with tradition and began to perform dissections himself, rather than assigning them to a surgeon. There was something of the showman in his irrepressible genius. He introduced "touch" and "feel" as indispensable aids to discovery and encouraged such personal engagement both in the practice of public dissection as well as in its visual representation. The youth was a thorough extrovert of unsubtle, constructive, and restless mind with an artist's eye, hands, and imagination. He understood the power of illustration and became the first to enhance his lectures by introducing large charts of the arteries and veins.

The immense popularity of those charts moved him to publish them. To his own three sketches of the heart, blood vessels, and viscera he added three plates of a skeleton that were drawn from actual specimens by Jan Stephan van Calcar, a pupil of the painter Titian, and he issued them all, without a title, as an atlas of anatomy. They were uncommonly large in dimension, 19 by 13.5 inches, with only one drawing per page. The power and novelty of that graphic publication, which became known as *Tabulae anatomicae sex* (*The six anatomical tables*), were obvious. It was neither a source for explanatory anatomical knowledge nor a work of artistic merit comparable to the earlier engravings of Dürer and Holbien or Leonardo's notebooks, but it did set a new standard, both in biological illustration and in the graphic arts of that day, especially in the illustrations of the skeletons.[3] Inferior, plagiarized copies, reflecting as they did some errors of the bookish Galenical tradition, appeared almost immediately north of the Alps at Marburg, Augsburg, Cologne, Frankfurt, and Paris. Only two complete original copies are known to exist today.

To supplement those "physiological diagrams," Vesalius brashly produced in the same year, most likely without his distinguished living teacher's permission, a new synopsis of the anatomical and physiological views of Galen that had been published earlier as a complete text by Guenther. And it must have been a great honor to the young academic when Agostino Gadaldino, chief editor of the prestigious Guinta (Junta) publishing house in Venice, commissioned him to undertake the Latin translations of three of Galen's anatomical writings for a forthcoming quintessential Latin edition of Galen's complete works. They included *The Nerves*, *The Arteries and Veins*, and *On Anatomical Procedures*. For Vesalius, that was truly a labor of love.

LATER IN LIFE, VESALIUS REGARDED HIS RESIDENCE AT PADUA HIS miracle year. Nearly all the chief works by which his scientific fame would live were composed or planned during that time. It was early in his career there, at the age of twenty-eight, that he produced his masterpiece—a Janus-faced volume that pretended to be highly conservative but that was, in fact, evolutionary. He called it *De humani corporis fabrica* (*On man's bodily works*), abbreviated to *De fabrica*.[4] It is hardly possible to exaggerate the effect of that volume on the revival of anatomy in Europe.

When he began that anatomy treatise, he could hardly claim much maturity on the subject. His experience with dissection was still limited, nor was he qualified to stand as arbiter between Galen and nature. He had put out no original research other than compiling notes from other texts or their editorial correction. But he made swift progress over the little more than three years of its writing when Contarini, an enlightened judge at the Paduan Criminal Court, became interested in Vesalius and turned over the corpses of executed criminals for his study. By the time Vesalius visited Bologna in 1540, he had performed more than twenty-six demonstrations on human cadavers. By his single-minded discipline,

Vesalius changed the face of education in anatomy by redirecting it from a brief annual event into a permanent and major subject of study.

It was only when *De fabrica* was well underway that he realized the full extent to which Galen had substituted animal structures in his descriptions of "human" anatomy. Vesalius compared an articulated human skeleton and the skeleton of an ape and proved, without doubt, that Galen's account of at least the bones had not been based on humans, and he guessed that Galen had never dissected a human body. It was an outrageous guess because it meant, if taken seriously, that *all* of Galen's anatomical descriptions were potentially erroneous, and the business of anatomy now was essentially to start over again.[5]

But, as later scholars have pointed out, his addiction to the Galenical texts prevented him from correcting some plain and obvious errors and inconsistencies, notwithstanding his own Prelude, which read "As for the accuracy of our work, believe me that no item is here which the Paduan students have not themselves confirmed at my demonstrations of this year." In the iconic illustration on the title page of *De fabrica*, he presented himself *with* the ancients, all engaged in the ancient practice of anatomy.

THE PUBLICATION OF *DE FABRICA* CREATED A STIR THROUGHOUT EUrope. Here was what an anatomical treatise in the new age ought to be. It revealed the divine majesty of God and envisaged the body as a manifestation of the Creator's wisdom and purpose. God had left His testimonies in the marvelous "workings" of the human body for man to contemplate, and young Vesalius's position as a supreme master of descriptive anatomy was now unassailable.

The text inculcated the principles of a new anatomical study more effectively than any other work hitherto published and brought back the pursuit of anatomy to the dissecting room. Its modernity stemmed from its emphasis on the living body as a functional organism. It was no longer a catalog of structures useful to physicians seeking the proper vein for bloodletting or to surgeons probing a battle wound. What was abandoned, too, was the traditional ancient Latin and medieval custom of listing the plain facts for each organ, such as size, shape, substance, location, and connections. The volume displayed 277 woodcuts of incredible beauty, accuracy, and detail and gave a systematic and illustrated survey of the whole human body, part by part, and layer by layer. It was a beautiful piece of bookmaking.

The illustrations were superior to any contemporary work. In true humanist tradition, they reflected the classical concept of life and extolled the intrinsic value and dignity of the human body as the microcosm of the universe. Like great poetry, the whole production was permeated by an awareness of the natural history of humanity grafted upon the study of nature. Another testament to the book's modernity was the invention of an elaborate system of cross-references between the text and illustrations (an index) that made the volume unique in the evolution of the printed book. It was remarked that in some of those innovations Vesalius had been anticipated, but by no one had he been forestalled in respect to them all, except Leonardo.

De fabrica was issued with a smaller volume called *Epitome* (*Abstract*), conceived as a companion and path leading to the highway of the major work. The title was misleading, for *Epitome* was nothing more than an offspring of the earlier *Tabulae sex* on a more magnificent scale. In contrast to *De fabrica*, *Epitome* had only fourteen folio pages, with eleven plates that were most likely also executed by Calcar. It was a triumph of condensation; it served at once as a brief manual of descriptive anatomy as well as an anatomical atlas. The text was clear and impartial, with classical style and phraseology, yet admitted new words and phrases as were necessary for systematizing a new terminology. Nothing was written about comparative anatomy. No authorities were cited, and all homologies and parallels, such as the comparison of the heart's shape to a pyramid, were omitted.

THE HEART, LUNGS, AND THE VASCULAR SYSTEM WERE DEPICTED IN *De fabrica* with a detail and accuracy never previously attained, other than by Leonardo. It was clear that Galen's sins had been both of omission and of commission. Dissatisfied with the Aristotelian system, Galen had set up a masterly model that, for all its seeming radicalism, had not needed to be true. It just had to be plausible in accordance with the dissection of animals and the limited observation offered by major human wounds.

Vesalius's originality was most clear in his discussion of pores in the ventricular septum that Galen had invented to filter coarse blood from the right heart chamber to be refined into light blood in the left chamber. It was at once evident that there were no such passages. The human septum, Vesalius wrote, was formed from the very thickest substance, and although there were conspicuous pits on both its sides, none penetrated from the right to the left ventricle.[6] Vesalius probed with a fine bristle and could not find even the most hidden channels. Indeed, he remarked caustically, he was forced to wonder at "the art of the Creator" by which blood passed from the right to the left chamber through pores that eluded human vision!

The motion of the heart and arteries was a cognate problem. He noticed that when the heart constricted and its apex receded from the base, the straight fibers relaxed, and the transverse fibers constricted. That resulted in a reduction of the internal cavities of the heart, which Vesalius proved, to his own satisfaction, by inserting through a hole at its apex a bunch of rushes that were tied together into the shape of a pyramid, with the tip pushed toward the base of the heart.

Vesalius had published his views on the origin of the arterial pulse five years earlier in 1538, when he had revised Guinther's text. He had expressed the same more clearly the following year in his published "Letter on Venesection," where he concluded, from direct vivisections, that the *contraction* of the heart, albeit passively as Aristotle had taught, caused "spirit" to pour out into the great artery. The arteries dilated when the heart contracted and its contents overflowed into them. He demonstrated the same findings at an animal vivisection in 1540 at Bologna, where he exposed the heart and a remote artery and then asked his students to feel the two simultaneously to determine their proper relationship.

By now, Vesalius had also improved his vivisection techniques for studying the living heart and arterial pulse. He admitted a more radical exposure of the thorax that made the movements of the heart more accessible to direct observation and that also allowed direct comparison with the aortic trunk rather than a peripheral artery. He combined that wide exposure with an intubation of the windpipe (trachea) and periodic artificial inflation of the lungs with bellows to prevent "suffocation" (the first published record of endotracheal intubation and artificial ventilation!). He showed that the lungs shrank when the chest was punctured and, when the chest had been laid wholly bare and immobile, an animal could still be kept alive through the timely use of bellows to ventilate the lungs through the tube in its windpipe. Under those circumstances, even with the thorax immobile, a heart which had almost stopped beating could be revived.

That first description of endotracheal intubation and artificial ventilation was a decisive experiment. It should have administered the death blow to the Greek dogma that the heart's expansion and contraction were dependent upon, and synchronized with, similar motions of the thorax. When the lungs were collapsed and flaccid for some time without ventilation, Vesalius observed a weak "undulant surging and vermicular motion" of the heart and a weakening of the arterial pulse; when the lungs were artificially reinflated, the pulse and the heart became strong and rapid again.

But the academic pressure of Galenism proved overwhelming. Vesalius had been an editor of Galen's works and was now hesitant to oppose the cumulative tradition of a thousand years. For all his insistence on originality in anatomy, he kept a close eye on Galen in physiology and, indeed, much of the unity of De fabrica was derived from following the master. He reiterated many of Galen's traditional errors such as the liver being the center and origin of the venous blood and that there was an ebb and flow of blood. His views on the pulse, too, remained orthodox and singularly unrevealing. It was inconceivable to have any other views. On what could they be based?

Notwithstanding, physiological issues *were* prominent in his mind. "The third [book]," Vesalius explained to Charles V in his lengthy and rather incoherent dedication, "comprises the close network of veins which carry to the muscles and bones and the other parts the ordinary blood by which they are nourished, and of arteries which control the mixture of innate heat and vital spirit." In a final brief chapter, he recorded the traditional doctrine that the pulmonary vein "holds the air" and conveys it "to the left sinus of the heart, where it is going to perfect the material of the vital spirit." The same vein returned to the lungs the sooty and unsuitable materials from the left heart as by-products of its activities. Great anatomist as he was, he failed to see that the pulmonary veins contained blood.

Recognizing that mistake would allow his student Realdo Colombo to stake claim to the discovery of the pulmonary circulation (we will address this in a later chapter).

VESALIUS'S FAILURE TO FORMULATE A GENUINELY NEW INTERPRETA-tion of physiology was less a desire for an alternative system than a dissatisfaction with an existing one. Although he showed Galen's error regarding the septal pores, he did not suggest any alternative explanation as to how blood could pass from the right to the left ventricle. He would have to put something wholly new in place to remedy Galen's erroneous one and make a natural harmonious relation of body function. And there was too little time in that crowded year at Padua for him to move from a description of structure to a more deliberate attack on function. There was no sustained music in Galen's physiology—only chance airs and fragments, broken harmonies, and scribbled cadences. For truth was the missing music. It would have required Vesalius to draw everything together using the truth to make it sing.

And that, Vesalius was reluctant to do. "I find no pleasure in pursuing these and many other matters at greater lengths," he wrote. Perhaps it was a failure of his century. He had neither the passion of Galileo to refine truth from error nor the vision of Kepler. He was all too aware that his bold attempt to demonstrate even some of the plain and visible truths of anatomy that were contrary to Galen were raising considerable eyebrows, and he was fearful of jeopardizing his rising academic career by taking on even greater risks. Religion had become involved in new ideas and discoveries that had to conform to the articles of faith.

Theological authority was strong and militant, and, in the fear and confusion of Luther's Reformation, there was utmost pressure upon scientific innovators to avoid direct conflict with classical doctrine or the Catholic Church. The church determined what was to be counted as knowledge; natural philosophy, of which anatomy was part, was still a religious business. The battle for and against Galen was as much the battle for and against the Catholic Church. And the church had heard the cry of the prophet stirring to set humanity free. It saw a future with intellect undermining faith, opinion superseding belief, the world shaking off the yoke of Rome. One power was about to succeed another power. Vesalius was a northerner in Catholic Europe, and the defenders of Galen who opposed him the most were also the most conspicuous among those who were defending the church against Protestantism. Vesalius would be berated for his impiety and lack of faith as much as for the damage he was doing to the Galenic tradition.[7]

thirteen

A Disinherited Mind

I have become my own problem.

—Augustine of Hippo

A NATOMY IN THE SIXTEENTH CENTURY WAS A DANGEROUS UN-
dertaking. It both attracted and repelled the talented. Rome transformed doubts into crimes; intolerance hindered the search for truth.
Any unfolding of new facts or interpretation could only be kept in accord with tradition by an exercise of considerable ingenuity. Undoubtedly, the depth and precision of knowledge about the human body in
De fabrica did provide an essential foundation for a more precise anatomy than Galen's. Vesalius had the human material to correct Galen's
other obvious mistakes, and he did correct Galen on the anatomy of the
bones, liver, bile ducts, uterus, and septum. He demonstrated that male
and female human skeletons had the same number of ribs, thus defying
the widespread belief, based on Genesis, that men came one rib short
because of the creation of Eve.

Friends urged him not to publish *De fabrica*. Be content with what has been discovered, they advised, and let our descendants pursue the truth after we are gone. God did not make humans to challenge the very world that He had created, and those who explored it were playing with fire—the earthly fire of the heretic's pyre as well as the eternal fire burning in Hell. His tenure would be questioned by the university faculty and even his publishers would be threatened with massive boycotts. *De fabrica* would destroy his prospects in life, what with the Counter-Reformation in full swing, the Jesuits and the Inquisition at the height of their power and influence, and the Dominicans doggedly sniffing for any whiff of heresy. After all, the only precedent for an attack on an ancient author in that age of humanism and Hellenistic respect for antiquity had taken place half a century earlier when Niccolò Leoniceno had accused Pliny of making mistakes.

But to the brash youth, nature's secrets seemed within his grasp. The worldview would have to conform to *his* fresh breakthroughs! Given his adventurous spirit and the exhilaration of discovery, Vesalius did not pause to reckon the cost. In fact, he argued with some sincerity that Galen would be proud to see him rectify some of his errors and omissions.

Time would prove his friends right. Not all were pleased with their young graduate's frontal assault on tradition and inertia. The university knew exactly where it stood, even if it meant standing still. The very nature of academia discouraged radical ideas. Academia, even then, preferred minor adjustments to existing theories. Any idea newly expressed stirred a professor's blood to hostility. Old men with a reputation to maintain were not tolerant of dictation from an erstwhile student.

WHEN *DE FABRICA* APPEARED, A MAELSTROM BROKE OUT. HOW COULD anyone cast doubts on the septal pores without upsetting the entire working model of the Galenic human heart and body?

Jacobus Sylvius, Vesalius's former teacher at Paris, wily and old now and behaving like a pre-incarnation of Machiavelli, relished the opportunity to manipulate and exploit. Five feet tall at most, thickset and squarely built with a bulging forehead and face marked by smallpox, his eyes had a calm and deadly gaze. His nose, which was thick, had a veined knob at its end that was popularly held by students to be a symbol of malice. In character, he could only be compared to a bridge troll. Pliant and smooth-spoken though he could be, Monsieur Sylvius had a nature hard as bronze and a temperament crude and parsimonious. To his pupils, he was cheap and stingy enough to well deserve the epitaph they had crafted for him: "Sylvius lies here, who never gave anything for nothing / Being dead, he even grieves that you read these lines for nothing."

Perhaps it was all about a sense of vulnerability regarding the damage the new book might do to Galenism. Disturbing allegations against Galen had been growing strong for some time. The fear that a "final blow" might be dealt—and worse, that it might be dealt by his own pupil—was unbearable.

Sylvius believed in evil as a plain fact in human beings. When the heart is prepared for evil, opportunity is seldom long wanting, and his reaction to his pupil was prompt and savage. Sylvius fought with his pen, calling his former scholar a madman ("Vesanus"), though he would have preferred to fight with the sword and end the matter on the spot. He dashed off an appeal for a royal order to impose sanctions against that Flemish upstart. He steeped his pen in gall:

> I implore His Majesty the Emperor to punish severely as he deserves, this monster, born and reared in his own home, this most pernicious exemplar of ignorance, ingratitude, arrogance and impiety: and to suppress him completely lest he poison the rest of Europe with his pestilential breath.[1]

In Padua, too, in the cut-throat world of academia, not the least active among his opponents was his own former pupil, that thorny sprig Realdus Columbus, who had an ax to grind and much cause for resentment—he had prematurely, and unwisely, made known his desire to get Vesalius's position at the university for himself.

At the same time, the renowned German botanist and physician Leonhard Fuchs was so impressed by *De fabrica* that he published a student edition "to prepare a pathway for the reader for an easier understanding of Vesalius [who] was divinely inspired."

IN SUCH A BELLIGERENT ENVIRONMENT, VESALIUS HAD THE COURAGE not to play the hero. He was no rebel defying authorities like his contemporary Paracelsus. If his findings differed from those of some past master, he voiced his disagreement simply because he had to report what his own eyes observed. Hadn't Galen himself admonished in his *On Anatomical Procedures* to "make it rather your serious endeavor not only to acquire accurate book knowledge of each bone but also to examine assiduously with your own eyes the human bones themselves"? Even Hippocrates had written that the art is difficult, and Galen had confessed that some great anatomists do miss a point or two. And Vesalius *had* quoted Galen innumerable times in worshipful respect.[2]

He gave new lectures at Padua and many demonstrations, some resoundingly successful and others less so, and offered to test the veracity of his statements in a public dissection. His everlasting self-justification over even the most trifling points made those demonstrations an endless apology. But there were some striking sounds in the apologia and in his acceptance of the "harshness" of discovery. He lectured again at Bologna to justify his findings, giving explicit directions to others on how to dissect to either verify his findings or come to independent conclusions, and then again demonstrated at Pisa where an enlightened Duke Cosimo offered him a professorship.

Such tokens of support weighed little when compared with the ridicule of his peers. Vesalius lacked the calm strength to accept criticism. His entire being, his inner life, seemed like a kind of absence, a variety of fugue. In the end, he suffered from the defects of his own qualities. The pride that had upheld him through the years of study now made him quick to take offense and slow to recognize the insensitivity, fear, and jealousy of his rivals. His senses were no longer objective, and his own brain seemed to conspire against him. He had with him other observations that he had not included in *De fabrica* and, in a paroxysm of frustration, he burned them all. And when the slow-speaking, stammering emperor with the weak adenoids and whining voice offered him a royal post accompanying his court through Europe, Vesalius ended a career in science so gloriously begun.

Henceforth, his years were that of a physician at the Habsburg court of Charles V, then in its full luxuriance, the last Gothic extravagance of the northern Renaissance. It was a life *asolare*, spent in the purposeless leisurely passing of time with thoughts effortless and monotonous. Gourmandism raises the specter of illness, and Vesalius was busied in many useful ways as he soothed His Majesty's gout (and perhaps malaria) and healed the maladies of the bon vivant courtiers and prelates, the colonial governors and the literati, who, at once aimless and intent, wandered like satellites around their emperor, indulging in the pastimes of princes and the games they proscribed and their Flemish pleasures of feasting and drinking, hunting and whoring.

THE MIDDLE OF THE CENTURY WAS IMMINENT, AND VESALIUS FELT A relieved sense that much of the old foul mischief that had been wrought was finally dying. He had lived in strange places surrounded by strangers speaking a strange language. There had been times when he had plunged into the desolation of shamed and terrified melancholia, devastated by the prospect of losing the higher faculties of his imagination as well as his intellect to mental illness. There was an inner uncertainty, too, which in one form or another offered his life a deep dilemma: a source of inspiration as well as confusion, now raising his thoughts to those heights to which only the force of tension could carry, then again trapping his genius in a tangle of insoluble contradictions. He had become the person he was so afraid of becoming: an agnostic—not even enough of a disbeliever to be an atheist. He wondered if he had betrayed the young man that he had been, as well as the God he had believed in.

Some ten years after its publication, his heart full of hope from a unique combination of wretchedness and high spirits, Vesalius set himself to the great task of revising *De fabrica* for a second edition. He decided to make changes predominantly in style and grammar. In deference to his loyalties to the Habsburgs and their strict Catholic observances, he resolved to omit all material that might be deemed objectionable to the orthodox. Remarkably, much of the credit that was given in the first edition to the work of his contemporaries was removed, perhaps reflecting Vesalius's obvious jealousy of the continuing accomplishments of others. He determined, in addition, that the new volume, with its heavier paper and larger type, should be far more sumptuous. The earlier letters would be recut and a new block prepared for the title page. Many improvements would be made, and many redundancies omitted.

A CHEERFUL LIGHT BROKE WHEN HE RECEIVED A LETTER AND A COPY of *Observationes anatomicae* (*Anatomical observations*) from Gabriello Fallopio, after whom the Fallopian tubes of the uterus are named. Fallopio had succeeded Vesalius at Padua. He had studied at Ferrara with Vesalius's friend Giovanni Batista Canano, and then proceeded to Pisa before assuming the chair at Padua. The author wrote to him warmly and at length and expressed an earnest desire to be regarded as a spiritual pupil of "the divine Vesalius." The majority of Paduan physicians, he wrote, still remembered and admired him.

Vesalius was overwhelmed. "Putting everything else on one side, he gave himself," as he said, "wholly up to the instant greedy reading of the pages." He was still supported as a person of consequence at the university! Reminded once again of his fulfilling year at Padua, Vesalius recognized in Fallopio a kindred spirit, a welcome light from the outer world. The universe of dissections and debates was out there, and he was here. And what lay between that past and the present, and had made him what he was now, was a wasteland, cold and desolate. Between the two worlds, mere words and figures on paper could not mediate.

As for the heart and vascular system, his own uneasiness about Galen's doctrine was becoming more manifest. Vesalius raged against his own inadequacies. There had been nights when he had awakened with a start, thinking he had heard the old man Galen laughing at him, goading him mockingly. He had suffered much. His face had become deeply careworn, as if every year had cut a new furrow on the brow, his body thin and wearied, his aspect that of a man who had made his final appeal for a desperate cause.

Putting those years of failure behind him now became his overruling passion. Better a shock of the new than indulge in old associations that mislead. He dipped his pen in ink. It bled:

> Not long ago I would not have dared to turn aside even a nail's breath from the opinion of Galen, the Prince of physicians. But the septum of the heart is as thick, dense, and compact as the rest of the heart. I do not, therefore, know in what way even the smallest particle can be transferred from the right to the left ventricle through the substance of that septum.[3]

HE HAD TURNED THIRTY-EIGHT AND WAS JUST AS ALONE AS HE HAD been when a child. Memories crowded in from the past. At one time, if he remembered well, his life had been a bowl of cherries where all hearts had opened and all wines had flowed, a gorgeous place that no dream could have contained. But now in the long winter evenings, Time itself seemed to burn like a candle without a wind, consuming the waxen minutes imperceptibly, his best years wasted, the precious attained only to be lost. He saw himself as a spent force, an activist who had outlived his moment in the anti-Galenic movement. He was, by his own admission, a failure. Wasn't his *De fabrica*, too, a process of progressive failure? In truth, he had failed, but his failure was more glorious than the successes of meaner men.

Shunned anatomist and retired academic as he had become, he felt once again the grist which had milled his inner life and by which he had given himself, Andreas Vesalius, a profound meaning. Admittedly, he had been arrogant in his youth, more daring than prudent, the bull in a china shop. He had taken false steps, blundered, exposed himself, offended in speech and word against tact and good sense. Such pain was not something one could purge but was something one lived with. He was living a shadow version of himself, with an awareness of how things might have been if he had done this and not that, if he had made this choice rather than that to raise his vendibility on the academic market. Pride asserts, humility testifies. And he *had* testified—many times, publicly at Bologna and Pisa, urging his detractors to either verify his findings or come to independent conclusions. But colleagues and critics had examined that talent with sterner eyes and ears and stomped upon it.

He had always remained an incorruptible fighter for academic freedom with unbribable integrity, which every artist and anatomist should insist are the necessary prerequisites. Every acquisition, every step forward in knowledge had been the result of courage, of severity to himself. Outside change had not brought a loss in inner steadfastness. And yet, though one can change one's situation and break away from objects and other beings, one cannot break away from oneself. Though the long years at court had passed, Vesalius still carried the torment from which he had hoped to escape. He could not change his ways by changing scenes. He had merely added remorse to regrets and suffering to sins. Now, Fallopio's book became a living voice from a brilliant world that Vesalius had once known so well, the last small bonfire of academic splendor that lay beyond the intellectual void of courtly life.

In 1555, the powerful Charles V, patron of Erasmus and the painters Dürer, Titian, and Tintoretto, astonished all Europe by abdicating and retiring to a remote monastery at Yuste in rural Spain. And there, veiled in decrepit age, he crossed the bar. Before his death, he granted Vesalius a pension and a pardon, removing the stain of illegitimacy from his family by royal edict. And he made him a count.

Eight years later, Vesalius's new patron, Philip II, the "dull, cold bigot of the Escorial," determined that dissecting human bodies was sinful. Born and bred a Spaniard under whom the great Armada was destroyed and Spain declined, Phillip was father of the ill-fated Don Carlos, immortalized by Schiller and Verdi. Now, he persuaded Vesalius to atone for his vivisections by undertaking a pilgrimage to the Holy Land.

On his way to Jerusalem, Vesalius stopped at Venice and renewed his contacts with his colleagues. There, he learned that Fallopio had died suddenly the previous year, and the senate had been at a loss for a fit successor. Perhaps it was possible that during his stay in the city, Vesalius, still attached to his colleagues by ties of intellectual sympathy, made known a willingness to abdicate his court title and return to academic life for the same position that Fabricius, Harvey's teacher, too, had applied. But that was not destined to be. On the way back from Jerusalem, amid one of the worst storms in living history, ill or shipwrecked, he was put ashore on the small, wooded, parrot-shaped Greek island of Zante, twenty miles south of Homer's rocky Ithaca, which had never known him as he had been.

Ancient Zacynthos had been founded by a hero of that name who was a cousin of Aeneas and a descendant of Dardanus. It was from here that the English poet Byron would leave for Missolonghi and to his death. And here, "in a vile and impoverished inn in a solitary place" remote from men and their fatiguing ways, a stranger in a strange land which was oddly innocent of culture and the rest of his world so far away, he passed his final season on earth. A wandering Venetian goldsmith who had happened upon the ailing Vesalius took his corpse and "with his own hands prepared his grave and buried the body so that it might not remain as food and nourishment for wild beasts." He was fifty years old.

In that same year, Michelangelo died, and Shakespeare and Galileo were born.

fourteen

The Medical Merchants of Venice

Anatomists and physiologists, I find you
wherever I look.

—CHARLES AUGUSTIN SAINTE-BEUVE

T HE RENAISSANCE WAS LATE IN PENETRATING THE SPLENDID ISO-
lation of the Most Serene Venetian Republic.[1] It is true that Giotto
had been in Padua painting the frescoes for the Capella Scrovegni. Do-
natello had been there, too, building the great equestrian statue of the
condottiere Gattamelata. Uccello had visited Venice to work on the
chapel at Saint Mark's, Dürer was in Venice twice, and Leonardo had
come in Giorgione's time. Jacopo Bellini, returning from a sojourn at
Florence, had opened his atelier at San Geminiano and offered lessons
in the "Florentine way." But it was more than a century after Giotto's
death that Jacopo's illegitimate son Giovanni, "manly John Bellini," as
Ruskin was fond of calling him, settled at Venice.

At that time, Venice was the wealthiest of all Italian cities and enjoyed a virtual monopoly on western Europe's trade with the East, with

which it was a vital link. Only during the severest winter storms did the gilded prows of galleys interrupt their coming and going. On the Rialto, or at Saint Mark's Square, could be seen people from every part of the civilized world, each dressed in their own way. Nowhere in all Italy could one live in greater liberty. Eventually, the city would become a synonym for magnificence and sophistication and for its aristocracy's glittering wealth. "Venetian splendor" in Renaissance Europe was luxury to disapprove of. And in huge processions from Saint Mark's under the auspices of the doge and the chief officers of state, the city could show how a great nation could rejoice at times of great triumph.

Economists most likely had Venice in mind when they claimed that "contemporary capitalism has invented nothing." It is true that everything was at Venice in embryo: bills of exchange, credit, minted coins, banks, forward selling, public finance, loans, capitalism, and colonialism. The Venetians invented the income tax, statistical science, the floating of government bonds, state censorship of books, and the ghetto. They were the first to make a regular income out of dreams by introducing the still-popular gambling casino and state lottery. It was a Venetian physician, Salamon, who had anticipated terrorism and biological warfare by concocting a plague-quintessence for use during the Turkish war.

HUMANISM ARRIVED IN THE CITY AND, WITH ITS ENHANCED REALITY, a reverence for the concrete world and the human body. When Giovanni Bellini was getting ready to paint the San Zaccaria Altarpiece at the height of his long career, a new *Anatomice* was published in Venice in 1502 by a Paduan professor of anatomy. Melanchthon called it the "thin and babyish little book." The author, Alessandro Benedetti, a man almost forgotten by historians, was the first to bring the general spirit of humanism to Venetian medicine by turning to pure Greek sources. He was temperamentally oriented toward the Greeks because he had traveled widely and had learned his Greek in Crete and in the Morea in southern Greece where he had lived for sixteen years.

A return to the Greeks, and an abandonment of detailed reference to the authority of the Arabs and the medieval Scholastics, was the hallmark of his new volume. A product of the Venetian High Renaissance, the work was well organized and clearly written, with many firsthand accounts of medical practice and anatomical anomalies. As a humanist and Hellenist, he argued for the superiority of Greek over Latin culture. Benedetti weeded out Arabic terms and self-consciously used Greek terminology. In a generation of natural philosophers, he might have been a great natural philosopher. But that wasn't enough. He was a true Renaissance man, well versed in Greek and Latin as well as in medicine, poetry, and philosophy. Among the audience at the well-attended anatomy courses he conducted at the height of his fame in 1500 was Leonardo da Vinci, who noted his own attendance in a notebook.

The heart was described in the third book of the *Anatomice* as the source and beginning of blood and the seat of nourishment and vital heat. Benedetti criticized at length the "shameless controversy" over the three chambers ("sinuses") of the heart that had been described by Aristotle. Admiration for Aristotle had become so disgracefully extravagant, he wrote, that anatomists were more apt to attribute their failure to locate a "third" chamber to their own want of skill rather than to any error made by the master. Some of Benedetti's contemporaries, indeed, identified only two main chambers as Galen had described, but elected to suppress what they saw and stooped to deceit to "save the phenomena"—a failure of nerve in deference to the weight of Aristotle's authority. Fear of what people may say is a curse in every order of things, even into our own times.

Benedetti wrote:

> But Aristotle has had so much authority for many centuries that even those things which the physicians have not seen they will affirm to exist, even without experiment. And if they judge they have seen only two sinuses, nevertheless they acknowledge that there are three in the heart. This matter is of great concern in all of medicine.[2]

BENEDETTI WAS HARDLY ALONE. THE MAGNETISM OF ANATOMY AT-
tracted many adventurers, each eager to beat out the other. Because
orthodox university professors would not touch any controversy in peril
of being ridiculed or losing their tenures, the field was open to those
who worked "outside the discipline."

Equally daring was his younger contemporary Niccolò Massa's introductory book on anatomy: *Liber introdutorius anatomiae*, published
in Venice in 1536, in which he described the prostate gland for the first
time. His first publication had been on the French disease, commonly
equated to modern-day syphilis. Massa was a libertine to the core. He had
no university appointment and was much of a freelancer, which made
him free to challenge authority while still dissecting corpses in the hospital of St. John and Paul in Venice. His book gave a remarkably clear,
though verbose, account of the human body, written by a skilled dissector
who was proud of his ability. It was a practical book. It outlined how to
see as much as possible during a single dissection, which made it popular with medical students. He criticized those who passed judgment on
anatomy without having applied a knife to the things they wrote about.

Massa knew the truth about the septum—the one truth! And it
was that the septum was very thick and dense and without pores. If
they had been perceived by men before him, he nevertheless now saw
their absence for himself with renewed and timely insight, and he was
bold enough to assert it.[3] Regretfully, Massa came across a congenital
opening, a "hole" in the heart, most likely what is now called a patent
foramen ovale. It is a communication in the septum not between the
two *ventricles*, which had been the focus of Aristotle and Galen, but
between the *auricles*. The normal physiology in the fetus requires a
communication, called the foramen ovale, to be present in the septum
between the two auricles to support fetal survival. That communication
invariably closes soon after birth. In some persons, it persists into adult
life as a persistent patent foramen ovale, abbreviated to PFO.

Notwithstanding, that solitary observation shook his belief, and he
muddied the waters further by equating that finding to Aristotle's "third
ventricle."

IT HAS BEEN USUAL TO BLAME "THE CHURCH" AND ITS GENERAL BAN on dissection for the virtual absence of progress in human anatomy before the Renaissance. The weakness of that argument is apparent by the appearance of a "new" treatise on anatomy by Mondino de Luzzi in 1316 in Bologna, a medieval city under tight church jurisdiction.[4] By the time of Vesalius, the liberation of anatomy had already begun. Even before Massa's introductory anatomy book, a short and well-illustrated *Isagogae brevis et exactissime in anatomis humanis corporis* (*A short introduction to human anatomy*) written by Jacopo Berengario da Carpi (Jacobus Berengarius Carpensis) had come out more than a decade earlier, in 1521.[5] It was an influential work written for "the common use of all good men," perhaps the most important book before Vesalius. It was, in fact, an abbreviation of Berengario's own nearly thousand-page survey of over a hundred ancient and contemporary authorities in anatomy that he had published separately. It widened anatomy as part of a broader movement for the understanding of humanity's place in God's creation and set a trail that would lead inexorably through Vesalius, Fallopio, and Fabricius to William Harvey.

Berengario was a prolific dissector with tremendous confidence in his own abilities. He was also a master of the *demonstratio*, whereby every proposition that he forwarded was "proven" by demonstrating the structure publicly, at dissection, to the sight and touch of his audience. He deliberately avoided "philosophy" in his anatomy by excluding those structures that could not be seen by the naked eye and those seen only by the mind's "eye of reason." He called his anatomy *anatomia sensibilis*, the anatomy of the perceptible, just as Massa had called his own an *anatomia sensata*, that is, an anatomy of what had actually been seen. Unique for his time (which was more than two decades before *De fabrica*), he used graphic figures to supplement the text when the complexity of a structure could not be easily described in words.

BERENGARIO WAS BORN IN THE LITTLE TOWN OF CARPI BETWEEN Modena and Mantua in northern Italy. Initiated into anatomy by his father Faustino, a barber surgeon, he obtained a doctorate from Bologna, where he studied under the famed anatomists Hieronymo Manfredi and Gabriel Zerbi, and was eventually appointed to its chair of surgery and anatomy.[6] He amassed a large clientele and a larger fortune, becoming physician to the rich and famous far outside Bologna. The Gonzaga of Mantua, the Estes of Ferrara, the Medici of Florence, and the pope in Rome were his patients, and to each he dedicated one of his books. Moreover, he strove to include prominent people in his audience during his dissections because "many eyes make proof," and if the eyes belonged to the rich and famous, so much the better. That had been a strategy that Galen himself had adopted when he had "proven" some of his opinions in demonstrations before the Roman aristocracy, who could then be named in his publications to add authority.

Berengario saw things as they truly were, and he was sufficiently pugnacious and socially and professionally secure to say so. He was among the first to declare firmly, even before Vesalius, that the pores in the septum did not exist in humans. He described the right and left sinuses (ventricles) of the heart as being separated by a dense, thick-walled diaphragm without perforations. He portrayed the heart valves accurately. He also denied the existence of a separate third ventricle as Aristotle and the Arab physician Avicenna had posited because he failed to see one:

> These orifices are considered by physicians as the middle sinus, according to Avicenna. . . . Avicenna, however, places a fourth part of the blood which he says is temperate in the middle ventricle. Yet this is unknown to my eyes . . . in that place there is no blood outside the veins, according to my judgment, except in the right and left ventricles.[7]

Clearly, in view of Benivieni, Massa, and Berengario, who all defied the septal pores as well as integrated illustrations in their anatomy books, Vesalius was neither entirely novel nor brave. But it was the landmark scientific, intellectual, and artistic quality of De fabrica in the history of the woodcut and the illustrated book, with its harmonious blending of art and anatomy, that moved the distinguished historian of science Charles Singer to remark that all of the former were picturing "dead Anatomy, Vesalius living."

The significance of the Venetian anatomists and Vesalius to our story is twofold. For the first time, fissures in the Galenic system were exposed, setting a preparation and precedent for Harvey's overturning of the whole of Galen's edifice a century later. Further, as in the case with Leonardo, the true knowledge at that time of any part or organ was based on its structure, on the accurate understanding of its anatomy, and in this, Vesalius revolutionized its study.

AIR AND BLOOD

Blood is a juice of a very special kind.
—Johann Wolfgang von Goethe

fifteen

Hunted Heretic

Science like life feeds on its own decay. New facts burn old rules; then newly developed concepts bind old and new together into a reconciling law.

—HENRY JAMES

WHEN VESALIUS WAS STUDYING AT PARIS UNIVERSITY (NOW known as the Sorbonne), it is doubtful whether he knew much about a colleague who called himself Villeneuve. Like most other students, he was aware that the youth was a favorite of their teacher, Johann Guenther von Andernach, who made him his assistant and considered him to be "distinguished by his literary acquirements of every kind, and scarcely second to any in his knowledge of Galenic doctrine." If anyone had any idea of the emotional turmoil of their classmate, he kept his suspicions to himself. Nor could anybody have guessed that Villeneuve was hiding his real Spanish name, Miguel Serveto de Villanova, otherwise known as Servetus.

Michael Servetus, alias Miguel Servet y Reves, brought together in a single person both revolution and reform.[1] He was born as Miguel Serveto Conesa on Saint Michael's Day at Villanueva de Sijena, a hamlet in the province of Huesca in Aragon at the northeastern corner of Spain. The house in which he was born still stands. He was a precocious child. When, at thirteen, he entered the University of Zaragoza and then the more prestigious one at Barcelona, he was already fluent in French, Greek, Latin, Spanish, and Hebrew, to which, when he turned seventeen, he added Arabic so that he could read the Koran. After a law degree from Toulouse, he changed his name to Michael Servetus. As the protégé of a Franciscan scholar, he traveled to Bologna, where he witnessed the embarrassing extravagance of the coronation of nineteen-year-old Charles I of Spain, "son of a seducer and a madwoman" and grandson of the Catholic monarch. As Charles V, he was the last emperor to be crowned by a pope. His titles included "Emperor of Spain, Germany, Austria, Naples, Sicily, the Low Countries and the New World, Defender of the Catholic Faith, and God's warrior vicar on earth."

DEPLORING THE SORDID CONFINES OF CONTEMPORARY SOCIAL LIFE, Servetus sought the consolations of theology. He was a physician by profession but an ecclesiastic by inclination and vocation. He strove to restore the simple words and meaning of biblical text against the artificial overgrowth of dogma that had obscured it over the centuries. His mind was entirely rational. He entertained no feelings, experiences, or convictions that he could not explain and justify rationally, including religion. He went to Basel, where he studied theology, residing as a guest of the preacher and reformer Johannes Oecolampadius, whom he found to be sympathetic to his own devout beliefs. He befriended a publisher, Johannes Setzer in Haguenau, a village near Strasburg. Through him, he brought out a pamphlet in 1531, fresh and full of spontaneity, called *De Trinitatis erroribus* (*Errors of the*

Trinity), a subject generally reserved for the most revered and erudite theologians. He was twenty-one.

The book was both an assault upon tradition and a reconstruction of his own position. The former forced him out of all Catholic lands; the latter made his stay untenable on Protestant soil. As his biographer (and biographer of Martin Luther) Roland Bainton pointed out, Michael Servetus had the singular distinction of being burned by the Catholics in effigy and by the Protestants in actuality. He soon realized that those who entrusted their soul solely to their beliefs in militant times invariably found themselves alone.

THE MOST SIGNIFICANT THEOLOGICAL UNORTHODOXY HE PROPOSED was his Unitarianism—he rejected the rather abstruse matter of the Trinity. He would become a guiding force, though not an actual founder, of the Unitarian Church. He held the Trinity doctrine to be a great stumbling block, and it conflicted transparently with the Hebrew affirmation "Hear, O Israel, the Lord thy God is One." He was amazed to find nothing whatsoever about a trinal unity when he delved into the sacred word in any edition of the Holy Text—Hebrew, Greek, or Latin. The expression itself was just not there, no matter how popular a notion it was. Nor was there any mention of the one substance and the three persons. "Not one word is found in the whole Bible about the Trinity nor about its persons, nor about the essence nor the unity of substance," he concluded. Even the key word, *homoousios*, describing the relation of the Son and the Father, was absent. So, if those doctrines were not biblical, when and where had they come from?

His studies brought to light that the Trinity dogma was verbalized at the Council of Nicaea in 325, three centuries after Christ, and formulated more precisely at Chalcedon in 450. It was a solution devised by theologians to posit both a oneness and a threeness in God: a diversity within unity, pluralism within monism. Luther had left that vexing problem out of his catechism, and many reformers had avoided the subject completely. Servetus, posing as Dr. Villeneuve, did himself few favors when he communicated his views in good faith to an ambitious but hitherto unsuccessful fellow student at Paris in the law faculty who was two years his senior: a French youth named Jean Chauvin, who had latinized his name while at the university to Johannus Calvinus. He would reinvent himself later as John Calvin. Calvin summarily rejected Servetus.

THE FOLLOWING YEAR, IN 1532, THE POLYMATH SERVETUS RELEASED another volume titled *Two Dialogues on the Trinity*, again published by Setzer. Is it any wonder that the magistrates forbade the sale of his work at Strasburg? In the same year, on June 17, the Inquisition passed a decree for the apprehension of some forty heretics. Topping the list was "Michel de Serveto alias Reves." Heresy was the supreme crime, even worse than holy matricide—the rape of that Immaculate Bride of Christ, the Holy Mother Church! For a while he considered the "New Isle" across the Atlantic as a possible asylum. He eventually emerged, like Edmond Dantès of *Monte Cristo* fame, somewhere in France under an assumed name, Michel de Villeneuve, finding bleak comfort in the king, Francis I, on whom religion sat lightly and whose attitude toward heresy fluctuated in accordance with his desire to form an alliance with the pope, the Protestant, or the infidel.

Servetus chose to hide at Lyon, where the great printing houses rivaled those of Basel in output and boldness. An upright French citizen now, "Villeneuve" supported himself as an editor at the eminent publishing firm of Trechsel. He edited Ptolemy's *Geography*, Paganini's Bible, and several works by Symphorien Champier, a medical humanist from Lyon. Maybe it was Symphorien who forced upon him the realization that a writer's existence with his publishers was a state of war. Perhaps, it was on his advice that he left editorship and set out to study medicine at Paris, where he worked with the cream of the faculty—Sylvius, Jean Fernel, and Guenther—and may even have sat on the same bench next to Vesalius.

A lonely man, never secure, he left Paris after graduation to practice medicine in the provinces. He was for brief periods in Italy, Switzerland, and Germany, and then at Lyon, Avignon, and Charlieu (where he may have had a romantic attachment with a local girl), pursuing the life of a wandering scholar, and finally settled at the ecclesiastical town of Vienne on the Rhone, about twenty miles south of Lyon, where he practiced for twelve years. It was a most idyllic life. Noblemen traveled great distances to consult him, and he acquired such a reputation that the archbishop made the famous "Dr. Villeneuve" his personal physician.

IN EARLY 1553, TWO YEARS BEFORE VESALIUS REVISED HIS DE FABRICA, Servetus published his thoughts anonymously in another book, *Christianismi restitutio (The restoration of Christianity)*.[2] It was a prodigious piece of scholarship, with references to over thirty sources in Latin, Greek, Hebrew, and Arabic. In it, he discussed, once again, everything that he opposed in the Bible: the injustice of infant baptism, the myth of the Trinity, and other contortions of Saint Paul and the Scriptures. The title was also a slap at Calvin's own book *Christianae religionis institutio*. Servetus sent a prepublication copy to Calvin with a lengthy letter. Calvin did not reply. Servetus persisted. The published correspondence between Servetus and Calvin would eventually fill a separate volume—one-sided, because Calvin neither replied to Servetus's thirty epistolary discourses nor did he return the manuscript copy of *Restitutio*. Servetus could never have guessed that it is but a short step for a reformist to turn reactionary.

Condemned by the church and denounced roundly by a now powerful Calvin, *Restitutio*—which also included all the thirty letters to Calvin—became, as a result, more widely read. At the same time, Calvin was informed, or rather misinformed, that Servetus was in league with the Libertines who were conspiring to overthrow Calvin and his regime at Geneva. Calvin now had two good reasons to eliminate Servetus: not only for the honor of God but also for the preservation of his own republic. Servetus was arrested by the Inquisition and charged with heresy but escaped from prison and took flight.

On Saturday, August 12, 1553, Servetus rode into the village of Louyset on the French side of Geneva, where he checked in at La Rose (the Inn of the Rose). He would take a boat across the lake to Lausanne. From there, he could link up with the Zurich road into Italy, which was more tolerant toward reformists and where he could practice medicine at Naples. The following day, having sold his horse, he went to an afternoon service. Church attendance was mandatory in Calvin's Geneva, and simply remaining at the inn would have been most suspicious, inviting inquiry. While attending church, his swarthy complexion betrayed him. He was recognized and denounced to the city authorities.

Why Servetus would knowingly return to Calvin's Geneva instead of taking several alternate routes into Italy will never be known. Such high tension cannot be sustained long, and when this one broke, the magnificent rebel was expeditiously presented a list of thirty-eight charges and judged a heretic. Calvin preferred to have him beheaded, but the magistrates sentenced him to be killed "as mercifully as possible, and without the shedding of blood."

By now his faith had been lived and forged into the very gestures of his body and had "conferred that unruffled confidence in life which enables one to go through the world without turbulence of heart." He took the sentence without a tremor and, with a certain indifference to circumstances, went to his execution with the air of one accepting a disagreeable, but unavoidable, invitation to dinner. On October 27, 1553, he was burned alive on the hillside at Champel, a suburb of Geneva, with a copy of his book tied to each arm. Witnesses observed that the burning was agonizingly slow, lasting for a full half hour, his ashes commingling with those of his book. It was hard for the spectators to put up with the screams.

History is invariably unjust to the vanquished. Like other reformers, Servetus played out his destiny to a fiery end. Monuments to his memory would be erected at Madrid (in 1876), Paris (1907), and Vienne (1910). Of the one thousand printed copies, virtually all were seized and destroyed, some with the author's effigy at Vienne and the rest with the author himself. Only three remain today, and only two are complete.

SERVETUS REGARDED A STUDY OF ANATOMY NECESSARY FOR A TRUE understanding of God. Like Melanchthon and his Lutheran followers at Wittenberg, he believed that anatomy revealed not only the structures, arrangements, and purpose of the body but also ways in which the activities of the Christian soul were mediated in thought, imagination, and the will. Such a theme was far too important to be left to physicians alone. Moreover, anatomy and spirituality were related in a variety of ways. After all, anatomy showed how fragile man was, how delicate his brain, and how easily damaged his veins. It was a perpetual reminder of the transitory nature of this life and of the judgment of God to come. Indeed, anatomy might even serve as a meditation on death.

It was therefore necessary for Servetus to blend anatomy into his theological works. Just as he had denied the supreme Trinity, so, too, he denied the Galenic triadic hierarchy of natural, vital, and animal spirits. He claimed that in all those parts there was the energy only of one and a single spirit. There were not two kinds of blood, either, as was traditionally differentiated by the natural and vital spirits. The vessels contained but only one blood because there was only one kind of spirit in blood. That was the vital spirit, also called the natural spirit, which was the same in all vessels because vessels, he held, eventually communicated through the ends of the arteries and veins.

The greatest fruit of his Parisian period had not been a degree in medicine but a discovery. The Hebraic books Genesis, Leviticus, and Deuteronomy were the starting points from which he raised his main argument: that the "soul itself is the blood." Indeed, one of the charges brought against him at his trial was that he held the soul to *be* the blood, which led logically to the heretical inference that the "immortal" soul must perish with the body. Servetus maintained that the soul had been breathed by God into the blood of the first man in precisely the same way as the ancient Greeks had pneuma pass from the surrounding air into the heart.

He held the spirit or *anima* to persist in blood, and that he attributed to its constant regeneration by the contact of blood with pneuma derived from air. Here was a proper reason to study anatomy: to explain where and how that contact took place. The soul, the Deity's lofty epitome of Creation, was not a stingy thing that took refuge in the little corners of the left heart chamber or the brain. Provision had to be made for the broadest possible contact between air and blood. Galen had not only failed to make such a provision but had also been ambiguous regarding the place where the conversion of blood into its bright red form took place and the extent to which air had access to blood.

ONE MIGHT WELL SCOFF AT THE SIMPLE NAIVETY OF SERVETUS'S idea, but it is difficult to disagree with the concept. The left heart chamber or the septum was clearly not a fit place for such a perfection, though it had been prescribed for the past thousand years. What little did "sweat through" from the right into the left ventricle could not possibly be of any significance in the vital process. Servetus retained Galen's view that blood was made in the liver and carried by veins to the organs—indeed, that was one reason why he was unable to discover a systemic circulation. However, he did believe that the vital spirit had its seat within the heart, and it passed from arteries to veins through peripheral communications between them.

His first correction of Galen, announced boldly between pages 169 and 173 of *Restitutio*, was the direct result of dissection, in which he had become skilled at Paris under Guenther. The transfer of blood from the right to the left ventricle, he observed, could not be by way of the septum, which he found was impermeable. Secondly, he noted that the "arterial vein" (pulmonary artery) leading from the right-sided heart chamber to the lung was just too large—nature would never have provided such a large vessel merely for the nutrition of the lungs. There just *had* to be another purpose for its very large size! That, he reasoned, must be to allow blood to flow in abundance from the heart to the lungs. The ample communications that, he speculated, existed in the lung between the arterial vein and the venous artery must also have a purpose. Everything seemed to point to an important event occurring in the lungs! And what else could that be other than the mixing of inspired spirit with the "subtle" blood that the pulmonary artery conveyed:

> That communication does not, however, as is generally believed, take place through the median wall [septum] of the heart, but by a signal artifice the subtle blood is driven by a long passage through the lungs. It is prepared by the lungs, is rendered yellow [light] and from the artery-like vein [pulmonary artery] is poured into the vein-like artery [pulmonary vein]. Then in the vein-like artery it is mixed with the inspired air.[3]

He concluded:

> And so, at length it is drawn in, a complete mixture, by the left ventricle through the Diastole, stuff fit to become the vital spirit. That the communication and preparation does take place in this way through the lungs is shewn by the manifold conjunction and communication of the artery-like vein with the vein-like artery.[4]

With a bona fide copious blood flow in the lungs, anatomy and theology were blended. Two anatomical observations, the absence of septal pores and the large size of the pulmonary artery, were enough for Servetus to infer that, between the "venous artery" and the "arterial vein," there was a mechanism in the lung to allow a broad interaction of air and blood.

Careful reading confirmed that Galen, too, had been familiar with a pulmonary transit, albeit as a limited trickle, and had believed in connections between the vessels in the lungs. In contrast, Servetus proposed a *bountiful* perfusion of blood from the right ventricle through the lungs to the left ventricle and arteries, particularly to the brain (we now know that as much blood passes through the lungs in unit time as through the entire body). From the "marvelous" action of the right chamber, and by a long passage through the lungs, blood became greatly agitated, prepared, improved, and rendered clear with a light red hue.

He explained:

> It is not simply air which is sent from the lungs to the heart by the venous artery, but a mixture of air and of blood, it is consequently in the lungs that this takes place. The light-colored flavus [hue] which characterizes the spirituous blood, is given to it by the lungs, and not by the heart.[5]

Servetus's description was monumental with potential far-reaching merit, although of no demonstrable physiological significance at his time. His stated objective was to locate the transfer of the Divine Spirit from air to blood "as is taught by God himself in Gen. 9, Levit. 7 and Deut. 12." In the process, he described the pulmonary circulation. In his mind, its significance lay not in the elucidation of blood movement but as a mechanism that explained the acquisition of the divine spirit in conformity with anatomically correct pathways.

Having inferred a new seat for blood and air admixture, Servetus was forced to refer the blood's color change to that new site because, according to Galen, the color was brought about by the blood's contact with air. "Just as by air God makes ruddy the blood, so does Christ cause the spirit to glow," is how he put it.[6] In his correct description of the pulmonary blood flow, Servetus was original and the first in the West to do so. However, he never grasped the idea of a true circulation through the whole body, maybe because he was not looking for one.

William Harvey, on the other hand, acknowledged the Italian Realdo Colombo as the discoverer of the pulmonary circulation, perhaps because he was unaware of Servetus's theological treatise. The knowledge of the pulmonary transit of blood was one of the necessary pieces of evidence that he had to possess before he could begin to solve the problem of the blood's movement throughout the body. That he knew of it by 1616 is evident because he copied the relevant passages into the notes for his anatomy lectures at the College of Physicians. Harvey also owed to Colombo the correct understanding of the contraction (systole) and relaxation (diastole) of the heart.

sixteen

A Eureka Moment

He dissects man, studies the play of his passions, inter-
rogates every fiber, analyses the whole organism. Like a
surgeon, he feels neither shame nor repugnance when
exploring our human sores. His only concern is for the
truth, and he lays out before us the corpse of our hearts.

—ÉMILE ZOLA

MATHEUS REALDUS COLUMBUS, OR REALDO COLOMBO, BORN in the medieval musical city of Cremona in Italy, was only a year younger than Vesalius. Of all the Italian anatomists, he was the most adventurous. He possessed an almost voluptuous intellectual energy and was largely unconstrained by the classicism of the day.[1] When Vesalius left Padua to supervise the publication of *De fabrica*, Colombo was appointed in his place. On the assumption that his master's departure was permanent, he promptly advanced his own reputation through unrestrained vitriolic attacks on the absent professor. At a dissection he conducted, he boasted that he was identifying anatomical structures

122

that the "great" Vesalius had failed to observe. The unexpected return of Vesalius must have been a real embarrassment. When Vesalius held a second farewell anatomy demonstration, Colombo was conspicuously absent. Upon Vesalius's final departure, Colombo again assumed his position for a year, then withdrew to Pisa to accept the new chair of anatomy that Duke Cosimo had established.

The scene changed again when in 1548 he moved to the Sapienza, the Papal University in Rome. By this time, that queen of cities had become a vast ruin, a huge compost heap of human hopes and ambitions. The Temple of Jupiter was a dunghill, vegetables were grown in the Circus Maximus, peasants lived and worked in neglected vineyards amid the broken rubble of the Palatine Hill, goats grazed wild on the Capitoline Hill, and the Forum was known as *il Campo Vaccino*, or "the Field of Cows," after its own herds of livestock. Every decade, the residents of the *caput mundi* (capital of the world) were decimated by epidemics of malaria or plague emerging from the foul-smelling, sewage-piled Tiber as well as the desolate stretches of the marshy Campagna that strayed up to the city walls. In this wicked, wolf-suckled city, Colombo found his proper place. Here, he lived the remainder of his life in pursuit of success, social status, and popular acclaim.

Eccentricity and a disregard for conventions were at no time unsympathetic to him. Among the pleasing errors of his young mind was his opinion of his own importance. He suffered from habitual and sustained self-reference and was never able to suppose that other people could feel differently. Everyone had a duty not only to fit in with his decisions but to applaud them. It was the gratification of vanity, which was with him as with many of us "the stimulus of toil and balm of repose." Whereas Vesalius had the deeper nature, the higher ideals, the more sensitive conscience, Colombo was too well amused by the follies of others to be concerned about his own.

Condescension turned to self-righteous anger and a sense of outrage when others failed to see things his way, especially when they failed to do so quickly. His presumption, his snobberies, his abrasive manner and arrogance, his maneuverings and manipulations, all mobilized critical detractors, especially other less talented anatomists upon whom he pushed his own views with an obstinacy and immodesty that often verged on bellicosity. Moreover, though completely intolerant and insensitive to the feelings of others, he was exceptionally thin-skinned himself and quick to take offense—not an uncommon association. Affable but treacherous in diplomacy, he recognized no morals and broke a promise with a light heart even as he was ever ready to make another.

But in the eyes of the Vatican and the grandees at Rome, he was an enfant terrible—quick-witted, brash and gay, gloriously full-bearded, strikingly unorthodox and avant-garde, destined for great things—which protected him from the controversies to which Vesalius and Servetus had been subjected. He was, as one might have guessed, no stranger to ambition. But he knew full well that ambitions could continue at Rome only through the privilege of Vatican benevolence or a powerful ancestry, both of which he lacked. And what ambition can be attained in any society without money?

At home with the ways of the scheming Guiliano Della Rovere, now Pope Julius II, and his rich, lively, and contentious papal court, it was not long before Colombo found his way into the cabals and boudoirs that abounded in that center of the world's spiritual life. Ever ready to fish in troubled waters, he was sure that in the event of success, he would get the lion's share of the catch. He threw borrowed money around, both in extravagance and benevolence in a way native to a man who, never in all his life, has exactly understood what money was. Like some of us, the mere making of money was not enough but being known to have made it; not the accomplishment of any great aim but being seen to have accomplished it. That thirst for applause, "the last infirmity of noble minds and the first infirmity of weak ones," was the strongest influence on his humanity. For what Colombo feared above all else was being counted as nothing. His temper was prone to the ardors of the flesh; he liked women but loved boys better, and it is uncertain whether he ever married, though he fathered sons.

In academics, as in love, it does not do to give oneself wholly. Colombo climbed the papal ladder by being lifted from above, ingratiating himself only as required by winning friends among influential people. Even the feeble and aging Michelangelo, a tired old man from whose soul many illusions had fallen by now, came once to watch Colombo demonstrate the muscles on the corpse of a servant boy. Earlier, when the artist had been working on a figure of the crucified Christ, he had dissected some corpses sent to him by the abbot of Santo Spirito, and it was rumored that Michelangelo and Colombo had once considered cooperating on a book on anatomy.

THERE IS NO PORTRAIT SO DARK AS TO BE WITHOUT ITS SOFTER SHADES. To that end, Colombo was not impractical; he dissected many more *living* animals than his contemporaries. Unlike Vesalius, he was intrigued by motion. He was concerned not so much with the shape of things or their arrangement or composition as by the movements they made. In one disagreeable experiment that he related with almost gleeful perversity, the chest of a living dog was swiftly opened with a knife and the heart grasped, summarily tied with a curved needle and thread, and then excised, throbbing and alive. Thereafter, the bonds by which the four legs of the creature had been secured were severed and the animal set on its feet. "And there is nothing more astonishing than to behold a dog *without a heart* barking and walking!" Colombo reminisced of that moment.

Vivisection allowed a study of motion. When the chest of a live animal was opened, the dilation and contraction of the heart could be clearly seen. Colombo recognized that it was easy to confuse which of those phases was contraction and which was dilation. The heart, he noted, could receive materials more easily and with less labor when the organ was relaxed and quiescent, whereas during ejection there was a need for greater strength. And so, he correctly documented that ejection took place during contraction. He observed that when the heart contracted the arteries dilated, and vice versa. In that way, blood flowed through the vessels and then ebbed back.

Colombo had an uncanny sense for experiment. Despite Aristotle's teaching that there was cold blood in the right ventricle and warm in the left one, Colombo noted, not much to his surprise, that blood in both ventricles was distinctly warm. And, despite all of Galen's teaching, when he opened the blood vessels between the lungs and the left heart chamber (the pulmonary veins), no air or sooty residues came out. Only bright scarlet blood. And that was a reproducible phenomenon.

He further noticed that, despite the positioning of only a bi-cusped valve between the left auricle and the left chamber of the heart (which Vesalius baptized the mitral valve since it resembled a bishop's miter), blood did not reflux back into the pulmonary vein. That being impossible, he concluded with certainty that the bright red blood must have entered the pulmonary vein only from the lung and could not have regurgitated backward from the left ventricle. The heart received the *spiritus vitalis*, responsible for coloring the blood red, *already formed* from the lungs.

He wrote:

> Between these ventricles there is placed the septum through which almost all authors think there is a way open from the right to the left ventricle; and according to them the blood is in the transit rendered thin by the generation of the vital spirits in order that the passage may take place more easily. But these make a great mistake, for the blood is carried by the artery-like vein [pulmonary artery] to the lung and being there made thin is brought back thence together with air by the vein-like artery [pulmonary vein] to the left ventricle of the heart. This fact no one has hitherto observed or recorded in writing: yet it may be most readily observed by anyone.[2]

Here was a commonsense seizing of essentials! One thing Colombo wants us never to doubt, and that is the excitement he always felt as he contemplated the core of any new discovery. The one passion Colombo does not explain away is his own. It is the passion to convince. At any rate, the preacher in him never flagged.

Gathering momentum, he continued:

I for my part hold a quite different view, namely that this vein-like artery was made to carry blood mixed with air from the lungs to the left ventricle of the heart. And this is not only most probable, but is actually the case; for if you examine not only dead bodies but also living animals, you will find this artery in all instances filled with blood, which by no manner of means would be the case if it were constructed to carry air forsooth and vapors.[3]

While pursuing the pleasures of a papal puppet, Colombo found time to cobble together a book. It turned out to be a shorter and inferior imitation of *De fabrica*, as many realized. Colombo's sole motive for writing his only grandiose and egotistically titled production was an eye to scientific glory. The desire to live in the thoughts of humankind is not peculiar to any age, but it was felt perhaps with unwonted intensity by the men of the Renaissance. He called it, grandly, *De re anatomica libri XV* (*The fifteen books on things anatomical*), issued posthumously and unillustrated, which claimed to "contradict all, both ancients and moderns."

It was not enough for him that his ideas would be influential; he wanted to be remembered as a man, a personality. Clearly, he was no yes-man: in the twenty pages outlining the pulmonary circuit, he criticized Galen six times, Aristotle on five occasions, and Vesalius, his master, six times, as much to belittle them as to raise his own prestige. But his entire work comprised mostly old concepts with one gigantic new one in their midst: his pulmonary circulation described in Book VII. Such is the nature of science, which builds its grand structure by adding fact to fact, concept to concept, tool to tool.

Colombo's flame burned itself out through its own intensity. We know that he died at Rome in the summer of 1559, at the age of forty-four.

seventeen

Arabian Knight

Life is short, the art is long. Opportunity is elusive,
experiment is perilous, and judgement is difficult.

— HIPPOCRATES

COLOMBO WAS MOST LIKELY AWARE OF THE WORK OF AN ARAB physician named Ibn al-Nafis (Abu-el-Hassan Ala-ud-Din Ali ibn Abi-el-Hazam, to give his full name) who, in the thirteenth century, had recorded the impermeability of the septum as well as described an alternate path for blood flow and aeration in the lungs.[1]

We know that Latin translations of Arabic authors were being edited during the Renaissance by the Alpago family of Veneto. Andrea Alpago, a traveling physician and scholar, had spent thirty years in Syria and its environs, where he served as physician to a Venetian colony at Damascus. He acquired a fluent knowledge of Arabic and rendered accurate translations into Latin of major Arabic medical works that had reproduced, and commented on, Galen. After he left Damascus, he resided for a few years at Nicosia in Cyprus, then returned to Italy and taught medicine at Padua till the end of his life. From the foreword to his 1521 translation of Avicenna's influential million-word medical textbook *al Qānūn* (the Canon), it was apparent that Andrea had read Ibn al-Nafis. It is possible that Andrea may even have translated that treatise in which Ibn al-Nafis had described the pulmonary circuit.[2]

Andrea's nephew Paolo, who shared his uncle's interest in the Arabs, was a medical student at Padua at the same time as Colombo was teaching there. And Paolo may well have mentioned to Colombo the idea of a pulmonary transit as had been conceived centuries earlier by Ibn al-Nafis. Another Latin translation of Avicenna, published in 1547 during Colombo's lifetime, had indeed included a translation of a "Commentary" by Ibn al-Nafis.

IBN AL-NAFIS WROTE VOLUMINOUSLY IN A STYLE PECULIARLY HIS OWN, not only on medicine but also on philosophy, religion, and literature. In addition to a comprehensive *Kitab al-Shamil* (*Book of medicine*), a separate volume on ophthalmology, and a treatise on the arterial pulse, he composed commentaries on Hippocrates, Galen's anatomy, and the Arab masters Hunayn and Avicenna, including a separate *Epitome* (*Abstract*) of the latter's huge *Canon*.[3] Indeed, Ibn al-Nafis's colleagues called him a second Avicenna. His therapeutics was simple, with a preference for dietary regimens over medicaments and simple drugs instead of complex mixtures.

To his shrewd mind, the teachings of Galen were fraught with imperfections. But he had carefully concealed the revolutionary wolf within him by donning, with great humility, the customary clothing of a Galenic sheep. After all, the officials of the caliph at Cairo should find nothing objectionable in the teachings of the head physician at the city's prestigious Al Mansouri Bimaristan (hospital; *bimar* means "sick" in Persian, and *stan* means "location"). It is still operational as an ophthalmology institute. In the introduction to his *Commentary on the Anatomy of Avicenna's Canon*, he humbly expressed that his aim was to pursue what had been illuminated in *al Qānün* and uphold the words of the revered Galen whose books he called the best.

Ibn al-Nafis agreed with Avicenna that the foundation of life was the heart, which he called "the sun of the microcosm." But Galen's invisible pores in the septum continued to vex him. The more he studied the records of Galen and Avicenna, the more his inductive logic led to a necessary conclusion: the septum "has no apparent opening . . . fit for the passage of this blood as Galen believes." Forbidden by the "authority of the law and our inherent compassion" to practice animal vivisection or dissection of human cadavers, he had to remain content with the study of animal carcasses, which confirmed that the partition of the heart, when cut, was no different from the walls of the ventricles themselves. If there were no pores, blood could not possibly penetrate the septum.

He wrote:

> Moreover, there is absolutely no opening between these two ventricles. . . . Anatomy defies what they say, for the septum between the two ventricles is much thicker than any other, preventing the passage of any of the blood and spirit, and thus their loss. Also, the statement that the septum is porous is wrong. What made him [Galen] accept this view was his belief that the blood in the left ventricle reaches it from the right ventricle through these pores. This belief is fallacious.[4]

Ibn al-Nafis held that a homogenous compound would not result if air were mixed with blood when it was still thick. The heart itself, with its usual contraction and relaxation, could not possibly retain the required blood consistency long enough to allow proper and complete admixture of air and blood within its left chamber. There just had to be some alternative location where blood could be rarefied sufficiently before it mixed with air to form the vital spirit. The right-sided heart chamber appeared admirably suited for that function. But how could rarefied blood be transferred to the left heart chamber where the vital spirit was created without passing through the septal pores, which he did not believe in? Ibn al-Nafis proposed a logical alternative. After all, as the Prophet had taught, when one is unable to pass through a place, one must pass either at the side of it or over it!

He wrote:

> After the blood has been rarefied in this [right-sided cavity] it must of necessity pass to the left cavity, where the animal spirit is generated. But there is no opening, as some thought there was between these two cavities, for the septum of the heart there is tight, without any apparent fenestrations in it. Nor, as held by Galen, would an invisible opening be suitable for the passage of this blood, for the pores of the heart are not patent and its septum is thick. The blood, therefore, after thinning passes through the vena-arterialis [pulmonary artery] to the lung for circulation and mixing within the pulmonary parenchyma. The aerated blood gets refined and passes through the arteria-venalis [pulmonary vein] to reach the left cavity of the two cavities of the heart, after having mixed with the air and become suitable for the evolution of the animal spirit.[5]

Both the deduction and the concept were evolutionary. With its abundance of air as well as proximity to the left heart chamber, the lung appeared to be a natural organ for the combination and "cooking" of a substance that could be fit for the generation and nourishment of the vital spirit. In the absence of septal pores, the portion of blood in the right heart chamber had no option but to pass through the "vena-arterialis" (pulmonary artery) to the lung. *The blood in its entirety made a necessary pulmonary transit through the lung.* Residual blood, left behind in the lung vessels, was used by that organ for its nutrition. Air that "is overheated and useless and not fit for the creation of the spirit" left over from the spirit-making reaction in the left ventricle "is carried back . . . to the lung."

Ibn al-Nafis's inferences were a milestone in the history of the circulation.[6] He was no experimentalist, and what he had to say depended on true observation and honest study of Galen's facts. After all, Galen knew that blood flowed from the right ventricle to the lungs and then trickled from the pulmonary artery into the pulmonary vein, but he had not developed that idea further in his voluminous writings. Ibn al-Nafis looked at the old scheme from a slightly different angle so that different facets caught the light and reflected it with a stronger beam. For the first time, the function of the lung was correctly defined as permitting admixture of blood with air.

THERE IS SOMETHING INCORRIGIBLY ROMANTIC ABOUT THIS GENTLE teacher and his quiet crusade against authority. But that unobtrusive Arab was destined for oblivion. He assiduously avoided publicity, like the Bedouin nomad who "receives his world as it comes from Allah and is not concerned to alter it more than he needs."

The closing years of his long life must have been happy. He received and enjoyed the respect and love of his students and colleagues.[7] Upon his eightieth year, he suddenly fell ill and took to bed. His physicians urged him to drink some wine, which they believed could improve his condition, but he refused as it was forbidden by his religion. A little more or a little less of life at his age did not seem to him a thing to make a fuss about. He awaited God's will. He knew of his impending death and bequeathed all his writings to the Al Mansouri Hospital, where they still reside. On the sixth day of his illness, on Friday, December 18, 1288, quietly and without much ado, he passed into an endless sleep. All Cairo followed his coffin to his grave.

eighteen

Who's on First?

*The more we treat the theories of our predecessors
as myths, the more inclined we shall be to treat
our own doctrines as dogmas.*

—J. H. THORNTON

IT HAS BEEN ARGUED THAT THE THREE PIONEERS OF THE PULMO-
nary circulation—Ibn al-Nafis, Servetus, and Colombo—may have
well begun their independent studies from the same literary source,
namely, the specific passage at the end of the tenth chapter of the sixth
book of Galen's *De usu partium* (*On the Usefulness of the Parts*). Galen had implied that a trickle of blood did pass through the lungs and
had suggested connections between the lungs' blood vessels. On the
other hand, Colombo's supporters pointed out that their master may
well have been unfamiliar with that Galenic source. Indeed, Colombo
not only failed to cite Galen in support of his discovery, which would
have surely eased its acceptance, but also presented it in strongly anti-
Galenic terms.

Even before Michael Servetus, other Europeans had made a few plausible and implausible claims. As is so often the case in the quest for truth, it was an instance when Servetus and Colombo came upon the same realization almost coincidentally, and there was no rivalry between them. Like Darwin and Wallace, or Leibniz and Newton, neither was probably aware of the other's work at the outset.

Regarding the priority of published announcement, there can be no issue.[1] Servetus announced his discovery in *Restitutio* in 1553. Colombo's *De re anatomica* did not appear in print until early 1560, despite the date 1559 on its title page. Since Servetus had been burned in October 1553, he could not possibly have known of Colombo's publication. Moreover, in the case of Servetus, it is possible to be even more exact. A manuscript from as early as 1546 revealed a passage on the pulmonary transit that was comparable to the later published version, thus establishing an even earlier priority.[2] On the other hand, Colombo's experimental demonstration might well have been less than original, given his probable familiarity with Ibn al-Nafis from Paolo Alpago.

THE CLOSE ASSOCIATION IN THE EARLY YEARS BETWEEN VESALIUS AND Colombo made it almost certain that the pupil was aware of his master's reservations regarding the septal pores. Colombo could not have failed to notice the alteration Vesalius made in the second edition of *De fabrica*. By that time, Colombo had most likely come to his own conclusions. Like Vesalius, he was bred in Galenic tradition, practiced the crafts of anatomist and surgeon, and was professor of anatomy and surgery. All knowledge that he possessed was also available to Vesalius; indeed, in intellect each was the other's complement. Vesalius was incomparably the riper scholar, the sounder critic, with a more reasoned judgment and a more cultivated taste. Colombo had the more fertile imagination, the brighter wit, and held the storyteller's inimitable gift. So how did Colombo get ahead of Vesalius?

In any field of discovery, observations do not emerge until we are prepared to see them. The facts had been there all the time for Vesalius, needing no new instruments or no new faculties. But the mind's eye was not in focus. Armed with just three facts, namely, the impermeability of the septum, the presence of only bright red blood and no air or sooty waste in the pulmonary vein, and the "one-way door" function of the mitral valve so that bright red blood could not enter the pulmonary vein backward from the left ventricle, Colombo demonstrated scientifically, as no one else had done, the admixture of air and blood in the lungs as well as a pulmonary circulation. His reasoning, based on vivisection and experiment, was systematically superior to anything antecedent.

On the strength of what his loyal pupil Juan Valverde of Hamusco—called Colombo's "Bulldog"—wrote, Colombo may well have been aware of the partial truth as early as 1556 (still three years *after* Servetus's publication). He had already, by that time, performed the vital experiments to demonstrate the presence of only bright red blood in the pulmonary veins and the competence of the mitral valve, as Valverde asserted in his own three-hundred-page *De la composicion del cuerpo humano* (*The composition of the human body*) published in 1556, which was within a year of Vesalius's revised *De fabrica*. Valverde wrote unequivocally of the absence of septal pores. The partition between the heart chambers was hard, thick, and strong like the rest of the heart substance, and although some sulci were present on its surface, no blood passed through them. Regrettably, Valverde took personal credit for the discovery of the pulmonary circulation by boasting that "nobody before me has said this."[3]

Servetus and Colombo both compromised with Galen to a lesser or greater degree. Servetus showed no striking departure from Galen other than his cogent theological reasons for the pulmonary circuit, and even retained the "fuliginous" waste vapors regurgitating into the pulmonary vein from the left ventricle because he had less understanding of the heart's valves. Colombo, on the other hand, realized that Galen could not be accommodated to his own findings at vivisection. He was ahead of Servetus and Vesalius in the completeness and directness by which he threw off Galen's yoke. But like many clever persons, he was misled by an excess of cleverness in the matter of the direction of the systemic *venous* blood flow—he maintained that just as arteries propelled the blood impetus from the heart to the organs, so did the veins *from the liver*. With *both* arteries and veins carrying blood outward toward the periphery, like Aristotle's well-worn irrigation system, there could be no question of a genuine circulation.

THERE MIGHT HAVE BEEN OTHER CONTENDERS FOR THE DISCOVERY of the pulmonary circuit. But those bids were either premature, exaggerated, or themselves swallowed up. The year 1559, when Colombo's book was posthumously published, saw another volume released, also in Rome, dealing with anatomical matters. Its author was Andreas Caesalpinus d'Arezzo, who, like Colombo, was a professor at the Sapienza and a favorite of Pope Julius II.[4]

In less than a dozen years, the young Caesalpinus had acquired a reputation as "the pope of philosophers." If Colombo lacked general culture, Caesalpinus drowned in it.[5] Learned in the wisdom of the pagans and an enthusiastic Aristotelian (most of his works carry the term *Peripatetic* in the title), he was abreast of all the new learning regarding medicine, philosophy, botany, zoology, anatomy, minerology, astronomy, astrology, and even demonology. He was not an experimenter but a theorist, even a disputer and activist against Galen and for Aristotle in all that related to medicine, and it was probably in that spirit that he enunciated his views. Some of his reasoning was thoroughly medieval.

In a digression in his *Questionum peripateticarum* (*Peripatetic questions*) published in the same year as Colombo's posthumous volume, he was quite clear regarding the pulmonary flow of blood, which he reiterated in his *Questionum medicarum* published in 1593:

> The passages of the heart are so arranged by Nature that from the
> vena cava a flow takes place into the right ventricle, whence the
> way is open into the lung. From the lung moreover there is another
> entrance into the left ventricle of the heart, from which then a way
> is open into the aorta artery, certain membranes being so placed at

the mouths of the vessels that they prevent return. Thus, there is a sort of perpetual movement from the vena cava through the heart and lungs into the aorta artery as I have explained in my *Peripatetic Questions*.[6]

Caesalpinus had been Colombo's student at Pisa and had probably heard him talk about the blood flow through the lungs instead of the septum. But as the title of his own work indicates, its whole purpose was to raise questions and argue them in the light of Aristotle's natural philosophy rather than enunciate new discoveries. That was its weakness as well as the reason for its ambiguities — traditional theories were juxtaposed with brilliant modern statements.

Caesalpinus mentioned, though only once and in passing, that blood was heated in the heart and transmitted from the right ventricle to the left "through the septum." But he was not convinced. On the same page, the novel idea of a pulmonary transit was put forward as counterargument, but both were left standing as such. In 1583, Caesalpinus published his best-known work *De plantis* (*On plants*), considered to be the first textbook on botany, in which his description of a pulmonary blood flow was repeated. And like Colombo, he had no idea of a general circulation, since he, too, still believed that blood ebbed and flowed in all vessels.

There was yet no inkling of a systemic circulation in the medical world.

nineteen

Wild Sea of Troubles

The heart of man is very much like the sea,
it has its storms; it has its tides.

—Vincent van Gogh

THERE COMES A TIME IN ALL THE SCIENCES WHEN IDEAS CEASE to bubble. Since the days of Galen's eminence, much had happened to spoil the élan of his system. Several men simply accepted the novel pulmonary circuit. Others felt compelled to come to grips with the problem of how the lungs could survive without receiving arterial blood and the vital spirits directly from the left ventricle. The real problem, however, was not that of septal pores or flows through lungs. The real problem was Galen's defenders, who stood ever ready to contradict Colombo to "save" another Galenic doctrine of doubtful authenticity. There are always so many more reasons for rejecting anything than accepting it. That such an end could be destructive is rarely realized.

Even a person who might have been a friend, admirer, and supporter of Colombo was amazed that a respected anatomist had, in fact, proposed that the vital spirit could be generated within the lungs. Gabriello Fallopio, former canon of the cathedral at Modena and teacher of anatomy at Ferrara and Pisa who now held the chair of surgery at Padua, ridiculed the whole idea during a public dissection held in the winter of 1560, even as Colombo's book was just publishing. Fallopio insisted that, unquestionably, the movement of the heart *must* be the cause of its heat, which, in turn, bred the vital spirits in that chamber. Additionally, Fallopio asserted, the observation that fish had no lungs and yet they moved and therefore had the vital spirit demonstrated unequivocally that the spirit could not be generated in the lungs.

On the other hand, Jean Guinther, the aging and influential Galenist and teacher of Vesalius and Servetus, supported Colombo's views in his own latest treatise, *De medicina vetri et nova* (*On medicine new and victorious*), published the same year as Colombo's posthumous work. It was not just unchanged air that came to the heart from the lungs to cool it like a fan, wrote Guinther, but rather a mixture of air and blood. Lungs refined blood and brought it together with air. Sylvius, surprisingly, who had opposed Vesalius so viciously, now reluctantly came to terms through his own dissections. In his new work, *Isagogae* (*Introductions*), published posthumously in 1555 and praised as "a systematic textbook of modified Galenic anatomy," he chose to remain silent over the issue of the septal passage.

Within a few years, both approaches were taught in Europe. Ambroise Paré, surgeon to four French monarchs, summarized the difficulties of deciding how blood was carried from the right to the left heart chamber. Paré himself acknowledged the pores but stated that they were not "perforated." He failed to understand that an essential feature of Colombo's discovery was as much the competence of the mitral valve and the presence of bright blood in the pulmonary veins as the pathway of blood through the lungs.

Both alternatives were presented by the Pisan anatomist Guido Guidi, known as Vidius, an acquaintance of Valverde and Colombo and professor at the Collège de France, who had read Valverde's book. Guidi confirmed from his own work that the septum did not have pores, and "therefore not even a drop of blood could pass directly from the right heart chamber to the left one." Moreover, he correctly focused on the key observation regarding the contents of the pulmonary veins: the presence of only bright red blood and no sooty residues, which, like Valverde, he had verified. Eventually, Guidi hedged. Regardless of what really occurred to blood, he said, the pulmonary veins, being soft in nature and having a single fine sheath, were certainly adapted for transporting *air* quickly and effectively from the lungs to the heart. Vidius is remembered today by the two structures that bear his name, the Vidian nerve and the Vidian canal.

Three years later, the Catholic Archangelo Piccolomini of Ferrara, personal physician to a number of popes, strongly rejected Colombo's thesis. He maintained that vital blood could only be formed in the left ventricle, and then reversed Colombo's own argument to show that, as a matter of theoretical necessity, blood in the pulmonary veins had to come *from* the heart to ensure the survival of the lungs. Andreas Laurentius, a prominent anatomist and Galenist at Montpellier, also "proved" that a pulmonary transit was impossible because any connections between arteries and veins in the lungs were still hypothetical. Galen, himself, had only speculated when he had said that the "vein-like arteries" took up a certain portion of the blood from the "artery-like veins" through subtle and invisible passages in the lungs.

THE CONTENTS OF THE PULMONARY VEINS (WHETHER AIR OR BLOOD) that communicate between the lungs and the left-sided chamber of the heart and the presence of connections between the terminal ends of the pulmonary artery and veins in the lungs were crucial to the discovery of the pulmonary circulation.[1]

Erasistratus in Alexandria had been among the first to teach that there was no blood in the pulmonary veins, only fresh pneuma from the lungs. Then Galen imagined a two-way gaseous flow: fresh pneuma from the lungs to the left heart chamber, and "sooty" wastes from the furnace of the left ventricle to the lungs for elimination. It was the unequivocal presence of bright red blood in them that spurred Colombo to study how it got there! It could not have come from the left ventricle because the proper function of the one-way leaflets ("doors") of the mitral valve guarding the entrance to that chamber would have prevented a reverse, or retrograde, blood flow through them. It could only have come from the lungs, with the mitral valve allowing flow unidirectionally from the lungs into the heart. The observation that the blood was bright red implied that the color change must have taken place in the lungs.

The "modern" history of the pulmonary veins begins with the Byzantine Oribasius, physician to the emperor Julian the Apostate in the fourth century.[2] Oribasius was among the first to place blood in them and suggest, erroneously, that the pulmonary veins carried blood *from* the left heart chamber to the lungs for their nourishment. The influential tenth-century Zarathushtrian physician Haly Abbas explicitly confirmed this. Haly's teaching entered the mainstream of medieval Galenic tradition through the physician-translator Constantine at the prominent medical school in Salerno. On the other hand, throughout their doctrines, the connections between the terminals of the pulmonary artery and pulmonary vein implied by Galen were ignored by other Byzantine and Arab interpreters. Avicenna, who shared the same beliefs as Haly, resurrected the anastomoses between the "vein-like artery" and the "arterial vein" with blood flow from the left ventricle to the lungs, which he, too, believed was necessary to "nourish" the lungs.

The speculated pulmonary anastomoses, however, were left out accidentally by Gerard of Cremona in his influential Latin translation of Avicenna's work, which then became the standard in medical schools. Galen's pulmonary anastomoses as well as Oribasius's presence of blood in the pulmonary veins, therefore, remained unknown for centuries in all subsequent Latin editions of Avicenna until they were restored in a revised translation of the *Canon* by Andrea Alpago in 1527.

MISTAKEN ARGUMENTS BASED UPON DISSECTIONS THAT DEMONstrated a congenital patent foramen ovale (PFO) continued to refute

Colombo. Just such a condition was encountered by Piedmontese anatomist Leonardo Bottalus, who contended that it was through that auricular opening that blood passed between the chambers in adults. The error was promptly rectified by Severinus Pinaeus, a learned Galenist, who correctly pointed out that the fetal auricular opening that Bottalus had observed could infrequently persist in the adult heart and, moreover, had been described as such by Galen. It was through *ventricular* septal pores, ascertained Pinaeus, that the Galenic transfer of blood occurred. Another PFO was reported by Giambattista Carcano, professor of anatomy at Pavia, in his book on embryology. Observations from the fetus were also extrapolated to the adult by Constantius Varolius, physician to Pope Gregory XIII. In fact, Varolius did not believe that it was necessary for any blood to go from the right to the left ventricle!

Giulio Cesare Aranzio, professor of anatomy at Bologna, had devoted his whole career to the anatomy of the fetus. He had described the fetal blood vessels called the *ductus arteriosu* and the *ductus venosu*, as well as the little nodules of cartilage in the semilunar valves that were eventually named after him as the *corpora Arantii*. He supported Colombo's explanation of the pulmonary transit in his own *Anatomicae observationes* in 1587. But he also allowed for a reverse flow of vital spirit from the left ventricle into the pulmonary veins: the vital spirits, he agreed, were generated in the lungs as Colombo had demonstrated, but they had to receive a supply of fresh blood in the reverse direction, too, from the heart.[3]

Occupying a unique position in the story of the pulmonary circulation was the splendid monograph *Anatomy of the Horse*, published posthumously in 1598 by Carlo Ruini of Bologna, who was not a physician or veterinary surgeon but a lawyer! The work did an equivalent service for equine anatomy as *De fabrica* had accomplished for the human body. It was the first book devoted to the anatomy of that noble beast, and its magnificent figures, and no less admirable text, were justifiably compared with those of Vesalius. Ruini gave a clear account of the pulmonary blood flow in the horse.

IN THE NORTHERN COUNTRIES, FELIX PLATTER, A PROMINENT SWISS physician at Basel, accepted the pulmonary circulation. The solidly Protestant Platter family and their rags-to-riches story were well known not only in the city but also farther afield, including in France. Thomas Platter the elder had fought his way through immense difficulties from illiteracy and poverty as a herdboy to a respected position as headmaster of a notable school and owner of properties, including a printing house. If that did not entitle him to fame, then it did sanction him to the bourgeois virtues of respectability and comfort. He looked forward passionately to the further advancement of his family through his elder son Felix, and Felix did not disappoint.

A shy young boy, earnest about his study as well as modest about his ability and ambition and conscious of the task ahead, he pursued medical studies at the Catholic school in the old Gothic town of Montpellier (which Riolan in Paris baptized the "modern Cnidus"), which still boasts pointed arches and ribbed vaults. He returned to his welcoming home five years later to set up practice and rose to prominence. He studied his own copy of *De fabrica*, which had been published in his hometown, and became an early supporter of Vesalius. Felix himself dissected more than three hundred cadavers.

On the other hand, Gaspard Bauhin remained uncertain. He was another influential professor at Basel upon whose demonstrations Harvey would base his own anatomical lectures. In his book, he described Galen's views regarding the septal pores back-to-back with Colombo's proposal regarding the pulmonary flow. Bauhin remained vexed over this issue for more than a decade. Eventually, in his *Theatrum anatomicum*, published in 1705 and considered to be the finest textbook on anatomy of that time, Bauhin could not help but acknowledge the presence of conspicuous pores, pits, and deep, narrow sinuses in the septum on the basis of an ox heart that he had boiled for a long time. It seemed apparent to him that blood must be carried through those pores that he had personally observed because nature never attempted anything rashly or in vain. Thereafter, he became less enchanted with a pulmonary transit but considered it proper to mention in a textbook.

In the Low Countries (the United Provinces), the famous Dutch anatomist Volcher Coiter also performed experiments and supported the pulmonary circuit. Having studied under Fallopio at Padua, Aranzio at Bologna, Eustachius at Rome, and Rondelet at Montpellier, he combined the lessons of his extensive education to become the first systematic exponent of comparative anatomy. Coiter examined the living hearts of cats, lizards, serpents, and eels and gave a detailed description of the beating heart of a newborn kitten. In terms of the systemic circulation, although he conducted ligature experiments and made observations similar to Harvey that when a vein was ligatured it swelled up on the side away from the heart, he failed to appreciate its significance as Harvey did. He did use the word *circulatio* but understood that to mean a to-and-fro movement, as in the rising and falling of fluid: evaporation followed by condensation.

PHYSIOLOGY OF CIRCLES

The motion of the blood may be called circular in the way that Aristotle says air and rain follow the circular motion of the stars.

—William Harvey

twenty

Prelude to Glory

No great discovery was ever made without a bold guess.

—ISAAC NEWTON

WILLIAM HARVEY WAS BORN IN A FORTUNATE LAND AT A FORTU-
nate time. One may ask whether he was a singular phenomenon,
an inexplicable figure of genius that appears only once every few centuries, or whether his arrival was in some sense more explicable. The
answer, of course, is both. Harvey was undoubtedly brilliant, but his
achievement would have been virtually unthinkable without the Elizabethan culture that surrounded him.

Even to a casual observer, those were heady times. Francis Bacon,
Galileo, and Kepler belonged to a generation only slightly older. Robert Fludd, Jan Baptist van Helmont, and Peter Paul Rubens were close
contemporaries, and Mersenne, Gassendi, Descartes, Bernini, Cer-
vantes, Shakespeare, and Thomas Hobbes were only slightly younger.
The fierce Counter-Reformation against Luther was in full cry. In 1618,
when the concept of the circulation was being defined in Harvey's mind,

the Thirty Years War had begun. Catholics and Protestants became embattled in a hot war in which the judgment was elementary—whoever is not on our side is a heretic, and the punishment for heresy is the stake.

It also remains debatable how far Elizabethan England may be called civilized. It was brutal, unscrupulous, and disorderly. To the godly who were contemplating a departure to the New World beyond the ocean, the devils were all too visible in a land that "aboundeth with murders, slaughters, incest, adulteries, whoredom, drunkenness, oppression, and pride." But if the first requisites of civilization are intellectual energy, freedom of mind, a sense of beauty, and a craving for immortality, then the age of Marlowe and Spenser, of Dowland and Byrd, was a kind of civilization.

Elizabethan lives reflected many of the qualities of that amazing age—its restlessness and impatience with old ways, its passionate enthusiasms, its eager curiosity and daring speculation in all fields, its boldness in action, and its abounding and apparently inexhaustible energies. London was overwhelmingly the beneficiary and location of most of those excesses. She was the world's arriving city of that day, a Mecca for enthusiasts and a place of refuge for exiles. Naturalists, astronomers, artists, scholars, adventurers, and charlatans converged there, many drawn from Protestant countries by the temptations of imperial patronage. With the passage of time, the city became impregnated with magic and tinged with melancholy and, in fact, ambiguous and ecumenical.

The great energies that gave their impulses to the spirit of Elizabeth's court were summed up by a youthful Shakespeare in the ditty "Some, to the wars, to try their fortune there / Some, to discover islands far away / Some, to the studious universities." The court mirrored the spirit of the age and was one of great confusion and contention, with two irreconcilable philosophies of religion engaged in mortal combat. Religious controversy had never been tolerant but was never less so than in Elizabethan England. Nominally Protestant, the Virgin Queen did not have the slightest regard for the proprieties and polite circumlocutions of a later day. In denouncing the Holy Father as the "whore at Rome," she meant just that! The pope was the Anti-Christ, which was how he was universally identified by the Anglicans.

What was the purpose of a Catholic Church anyway? The Puritan poet John Milton referred to it as the "beast in ambush on the Seven Hills" and described the pope as "the Latin monster with its triple crown [who] gnashed its teeth and wagged its ten horns with menace horrid." The conflict that centered on a fiercely contested right to freedom of conscience was merely one aspect of the still-larger right to freedom of thought and speech that floated the *Mayflower*. The battle shifted from the pulpit to the field, with ecclesiastical concepts so abstruse and nebulous that anything and everything could be read into them, and anything and everything was. With one accord, princes and prelates promoted or stemmed the rising tide by the most ruthless means, and with the usual lack of results in the long run.

Such was the background for William Shakespeare and William Harvey.

Born in Folkestone of Kentish yeoman stock on the first day of April, thirty-five years after the publication of Vesalius's *De fabrica*, and speaking with the Kentish twang, Harvey entered King's School at Canterbury in the same summer that saw the defeat of the Spanish Armada.[1] He was admitted to Gonville and Caius College, Cambridge, at sixteen. After graduation, which was held at the University Church of Great St. Mary in the heart of Cambridge, Harvey journeyed to Padua, then the city of new ideas and the Mecca of science. A new medical school had been erected on the western islet of Padua in the ancient heart of the city called the Piazza della Sapienza. He had heard many students refer to the school as *il Bo*, "the Ox."[2] The city welcomed tuition-paying foreign students, not least for economic reasons.

In an Italy hounded by heresy, Padua was an island of tolerance. It flourished because the doges at Venice knew how to defend it from the desires of both East and West, that is, through the glory of universities rather than by force of arms. Among Padua's distinguished alumni had been Nicolai Copernicus. In the events leading to Harvey's discovery, scholars trained at Padua made the greatest impact: Vesalius, Colombo, and Fabricius; and in the heated controversy that followed, graduates from Padua outnumbered those from other universities combined. Himself at one time a medical student, Galileo taught mathematics, astronomy, and astrology in the Faculty of Medicine. After all, doctors still needed to discern what the stars foretold of illnesses as well as divine the most propitious times for mixing medications. There, Galileo devised a machine to raise water, designed the first thermometer, and constructed one of the first "spyglasses" to view the heavens; his friend, the Greek mathematician Giovanni Demisiani, proposed the name "telescope" for the spyglass. The "microscope" was a by-product of his invention of the telescope. Galileo took in boarders at his home on the Via Vignali, a narrow, winding street near an open square, and gave private tuitions to students to augment his meager academic salary.

Harvey often slipped into the crowded amphitheater when Galileo lectured, and he learned to restrict his research to problems that could be solved by experiment and measurement. He enjoyed, too, the evenings at the Angel Pharmacy, more a scholars' club than an apothecary shop, whose owner, Giovanni Vincenzo Pinelli, possessed what had seemed the most wonderful library in Europe. No one could have guessed that the aged Galileo would be assigned forced residence by the Inquisition, as young John Milton reminds us in *Areopagitica*, "for thinking in astronomy otherwise than the Franciscan and Dominican licensers thought."

The tradition of Latin Aristotelianism, driven by the Italian philosopher Pietro Pomponazzi, culminated at Padua in the last and greatest of the traditional teachers, Jacopo Giacomo Zabarella, from whose book *De methodis* (*On method*), published in 1578, Harvey learned Aristotelian methods. Zabarella taught that there were two approaches by which a physician could study the human body.[3] First was the demonstrative method of science, as exemplified in Aristotle's *History of Animals*, whereby learning came through anatomical observations and empirical knowledge, without necessarily understanding their rationale.

The second was the analytical or resolutive method, whereby reasons behind the observations were sought and understood. That was embodied in Aristotle's *Parts of Animals*, in which causal explanations were offered for the observed anatomy. In medicine, the resolutive method could proceed from diagnosis to cures.[4]

RETURNING IN 1602 TO AN OVERCROWDED LONDON "PAVED WITH men," the English student began the arduous climb up the academic and professional ladder. The Oxford biographer John Aubrey has given us a good description in his *Brief Lives*: "He was not tall; but of the lowest stature, round faced, olivaster [olive-colored] complexion; little eye, round, very black, full of spirit; his hair was black as raven, but quite white 20 years before he died . . . he was much and often troubled with gout."

In 1604, Harvey married Elizabeth Browne, the daughter of a prominent London physician. They had no children, and little is known about her. In 1607, with Dr. Browne's influence, he was admitted as fellow of the College of Physicians, later to become the Royal College. The college had been founded in 1518 during the reign of Henry VIII at the instance of Thomas Linacre, Physician in Ordinary to the king, who then became its first president. Dissection was instituted in 1565 after a grant authorized by Elizabeth I. In 1609, Harvey became physician to Saint Bartholomew's Hospital on the death of his senior, Dr. Wilkinson, and in 1615 was named Lumleian lecturer in anatomy at the college. The Lumleian Lectures had been inaugurated in 1582 and named after Lord Lumley, one of the two benefactors of the college.

The wide scope of Harvey's reading was soon apparent. Aristotle was his supreme authority, but Galen, too, was called upon, and the same charity was accorded to his own contemporaries.[5] Equally impressive was his range of knowledge in comparative anatomy and pathology. He was an omnivore, and not an exclusively anatomical one given his taste for food and drink that were now beginning to show! He mentioned no less than 128 species of animal life in his lecture notes, including snails, slugs, crabs, shrimp, wasps, hornets, and flies, on many of which he had made personal observations and dissections. Dissection of the body was, for Harvey, always a wonder that kindled enthusiasm—the Greek word *enthusiasmos* meant "possessed by a god," for there was in the word the root *theos*, which meant "god." Dissection did not merely instruct about the vital functions but also enabled one, in a modest way, to approach the great secret of the Creation.

HARVEY WAS CALLED UPON TO ATTEND ON JAMES I, AND WITH CHARLES I he came to be on terms of intimacy. Sir Francis Bacon was his patient, though it is unlikely that they ever discussed the circulation because they held opposing views on Aristotle's philosophy. He may have known William Gilbert, the pioneer of magnetism, who, in the opening sections of his *De mundo* (*On the world*), explicitly criticized those who slavishly followed Aristotle and Galen. He may even have crossed paths with a young Flemish chemist, Jan Baptist van Helmont, who visited the court at Whitehall in the very year of Harvey's return to England.

The king placed the resources at Windsor and Hampton Court, with their parks and wide and exotic range of animals, at Harvey's disposal for the purposes of experiment. Harvey dissected innumerable royal deer since Charles was an inveterate hunter—the buck in the summer and the doe in fall and winter. He was even allowed an ostrich from the royal menagerie. In return, Harvey showed His Majesty the development of the chick in an egg and the pulsation of its living heart. When civil war broke out against Oliver Cromwell, the king entrusted his two sons, Prince Charles and Prince James, to Harvey, who was a devoted monarchist. Friendship was, for Harvey, a cult. Courtesy, too, was preeminent in his life, and he abhorred those who hurt people's feelings to placate their own ego. That Jeffersonian phrase, "a decent respect for the opinions of mankind," applied well to his manner of life.

From late 1638, and throughout the civil war, Charles traveled almost incessantly, and Harvey accompanied him everywhere, including those happier years at Oxford, where he was warden of Merton College. He attended the king during his captivity at Newcastle and, after Charles's shocking trial and execution, returned to London in 1647 to live with his brother Eliab. But the turbulence of war and its terrible aftermath left Harvey, then over seventy years old, quite undaunted. In 1649, he published his two *Exercises* in response to objections raised on the circulation by the French anatomist Jean Riolan (discussed in a later chapter), and then his voluminous classic *On Generation* appeared.

The record of his last years when he retired to London into private life must have a certain sadness, but it would be a great mistake to represent the end of Harvey's life as a period of gloom. On the contrary, his was a chronicle of the triumph of character over circumstance. Events, to him, were, after all, only the raw material of life; it's a man's way of dealing with them that makes or mars the finished article. To be sure, he had troubles enough to breed despondency, almost to excuse moroseness. He was assailed by attacks of gout, kidney stones, lassitude, and insomnia. Only toward the end did time seem to drag, with days empty and nights interminable.[6] But he had a scholar's tenacity, a scholar's courage, and a scholar's inexhaustible consolations.

His end came on the evening of June 3, 1657, aged seventy-nine, following a stroke in Eliab's house at Roehampton, then a charming rural retreat in the Surrey countryside. Aubrey continues: "The morning of his death about ten a clock, he went to speak, and found he had the dead palsy in his tongue; then he saw what was to become of him, he knew there was then no hope of his recovery; . . . he made sign to Sambroke his apothecary in Blackfriars to let blood from the tongue, which did little or no good; and so, he ended his days."

Having no heirs—his twin brothers Mathew and Michael had died in 1643, his second brother John in 1645, and his wife shortly after—he bequeathed his estate of 20,000 pounds, as he had earlier his entire library, to the Royal College of Physicians, directing them to use the proceeds to study the secrets of nature. Thomas Hobbes was willed 10 pounds "as a token of his love." He was buried in the outer chapel of the Harvey family tomb at Hempstead in Essex (still a simple village of 450 souls) between the bodies of his two nieces. Here he would lie unvisited for two centuries. Many efforts to convey his remains to a fit resting place at Westminster Abbey would come to naught. In 1847, the local villagers, when asked, knew by hearsay that "Dr. Harvey was a very great man, who had made some great discovery," though they did not know what it was.[7]

Alas, W. H.—we hardly knew ye!

IN THE ONLY RECORDED CONVERSATION THAT THE AGING HARVEY had with young Robert Boyle, he confided that the most significant observation of his career occurred during his student years at Padua, when his venerable old teacher Hieronymus Fabricius had demonstrated the valves in the veins.[8] It was one of those moments when science, helping to unravel itself, drops a clue. "He answered me," Boyle recalled in his notes, "that when he took notice that the Valves in the Veins were so placed that they gave free passage to the Blood towards the Heart but opposed the passage of the Venal Blood the Contrary way . . . he was invited to imagine that [Blood] should be Sent through the Arteries and Return through the Veins."

The venous valves had been described anatomically by Vesalius in the second edition of De fabrica, but his deductions were faulty. Some years earlier, the French anatomist Charles Estienne had also observed similar structures and interpreted them as "a sort of baffles" to limit blood from passing too fast in certain vital tissues. Vesalius first became aware of "certain peculiarities of shape in some veins" in the midsummer of 1546, when he was called to Ratisbon to attend upon Lord Francesco d'Estes's illness. There he met Giovanni Battista Canano, an anatomist from Ferrara, and his pupil, the Portuguese anatomist Amatus Lusitanus.[9] The pair had demonstrated the valves, which they called opercula, during their lectures and dissections at the University of Ferrara. They resembled the ones at the orifices of the aorta and pulmonary artery in the heart and were present at the commencement of the vena azygos which entered the vena cava, as well as in the veins entering the kidneys and those near the upper region of the sacrum. Both anatomists asserted, correctly, that their function was to prevent reflux of blood.

Canano and Amatus repeated their observations the following year. Canano, among the first to discover the valves, never published his findings. He did begin a book on muscles but abandoned it, perhaps discouraged by the appearance of *De fabrica*. Amatus, a Jewish physician from Castelo Branco who eventually became professor of medicine at Ferrara, described them in his *Centuria* (*Hundred observations*), published at Florence in 1551. He had noticed them in the azygos vein "a thousand times." Contrary to others, he demonstrated, by a clever experiment, that they facilitated a one-way flow of blood toward the heart. Colombo gave good descriptions too, but Bartolomeo Eustachius insisted that such valves could not be found. It would fall upon Hieronymus Fabricius ab Aquapendente (to give his full name) to rediscover them in 1574, five years before the death of Canano.

A MAN ENTIRELY SILENT AND TRANQUIL, HAPPILY MARRIED BUT CHILD-less, physician to Galileo as well as the Senate at Venice and the King of Poland, Fabricius succeeded Fallopio to the chair of surgery and anatomy at Padua. The valves in the veins were his obsession. He had noticed them as early as 1574 and demonstrated them publicly in his lectures. The Jewish physician Salomen Alberti, who attended Fabricius's demonstrations, published a report at Nuremberg, as also Piccolomini of Ferrara who had been Canano's pupil. In his anatomy book published in 1586, Piccolomini correctly described the function of the valves of the jugular vein as well as of the veins of the extremities.[10]

For most of his long career that spanned over half a century, Fabricius taught anatomy at Padua. To facilitate his work, he constructed an astonishing anatomy theater that can still be visited, the first permanent structure of its kind, where students flocked from all over Europe for demonstrations by torchlight. On a central dissecting table at the bottom of five circular tiers of seats, a corpse, the anatomical model, showed the students what became of humankind at the end of its earthly history.

Empty lives, thought Harvey; lives of uncertain appeal surrounded by glittering ostentation. But all the splendor in the world could not efface the naked body under the habit, the flesh perishable on the bones. The cadaver was raised through a trapdoor from a corridor below, and the dissected remnants were burned in a furnace. Fabricius correlated function, pathology, and therapeutics because he was not only an anatomist but also a physiologist and embryologist.

The definitive work on the valves appeared in 1603 when Fabricius published his illustrated *De venarum ostiolis* (*On the little doors in the veins*).[11] While in his late sixties (he would die aged eighty-seven), Fabricius conceived an ambitious book project, titled *A Theater of the Whole Animal's Workings*, in which, like his other anatomy theater, he planned to exhibit and display the entire business of anatomy and physiology to an international audience. He did not live to complete such a monumental task. But he did manage to put out a series of monographs intended as contributions toward it. The treatise on the venous "doors" was one such component.

Fabricius was the first to use the word *valves* to describe those delicate little membranes, which resembled fingernails attached to the walls and opened upward in the direction of the main stem of the veins. They were fragile and occurred at intervals, singly or in pairs, and most marked in the limb vessels. They were easily noticeable through the skin in the arms of thin folk when a ligature was applied, at which time they resembled the swellings seen at the nodes of the stems of plants. All veins did not have valves. They were absent, for instance, in the vena cava, the *internal* jugular vein, and those supplying organs that demanded an uninterrupted blood flow, as well as countless small superficial veins.

Despite his accurate description and his belief that "nature had not placed so many valves without design," Fabricius failed to appreciate their true design, which was to allow blood to return *from* the periphery *to* the heart. He still believed in a Homeric ebb and flow. In fact, he demonstrated the Galenic origin of the veins in the liver as well as a passage in the septum between the two ventricles by means of a probe! The function of the venous valves, he maintained, was to strengthen the veins as well as retard flow to enable each body part, or organ, to allow enough time for a complete "concoction." Additionally, they prevented overdistension of the veins.

There was a reason, he inferred, why the *internal* jugular vein was devoid of valves—here, blood could brook no delay since the brain, as one of the principal organs, needed an unchecked supply of fresh nourishment (veins were still regarded as carrying blood *to* organs). Likewise, veins to the liver, heart, lungs, and reproductive organs had no valves. They were all essential for the conservation of the whole animal and therefore required a most copious supply of blood. Arteries did not require valves because, in the traditional system, they were not as concerned with nourishment and, given the strength and thickness of their coats, did not need reinforcement either. It would be left to Harvey to demonstrate what their function really was and, together with other findings, reexamine and reevaluate all theories concerning blood flow.

FABRICIUS'S EXPOSITION OF THE VALVES IN THE VEINS WAS PIVOTAL TO Harvey's discovery. Fabricius had unwittingly given him an essential piece to the puzzle of his own research. Aristotle and Galen, and after them all anatomists including Fabricius, had taught that blood went to the organs and extremities and did not return. For fifteen centuries, the whole of medicine had rested upon that centrifugal system away from the heart. Yet here were the very structures that could overthrow the tyranny of the Galenic ebb and flow of blood. They were the obvious proof that blood must return from the periphery to the center through the veins. And Fabricius had failed to appreciate the enormity of his own finding, leaving Harvey the good fortune to complete the truth: blood flowed in the body as a unidirectional circulation!

twenty-one

The Finest Hour

Thus, Harvey sought for truth in truths own Book
The creatures, which by God himself was writ.
And wisely thought 'twas fit,
Not to read Comments only upon it,
But on th' original itself to look.

—Abraham Cowley

HARVEY'S NOVEL DISCOVERY THAT "THE BLOOD TRAVELED ALONG a previously unrecognized circular pathway of its own" was the outcome of long and patient labor, amassing a wealth of evidence from vivisection, visual observation, and experiment, especially with ligatures.[1] His findings were based on an empirical study of more than a hundred different species, including mammals, serpents (snakes), fish, lobsters, toads, lizards, slugs, and insects, which led to a radical reform of the physiological basis of medicine.

Several novel aspects to his contribution brought the life sciences into the modern era as full-fledged participants in the scientific

158

revolution.[2] First, Galen had viewed the veins and the arteries as wholly independent systems. Harvey introduced the concept of a single continuous system in which arterial and venous blood were one and the same.[3] Even if arterial blood contained more spirits, it was still blood, as the spirits were not separable but were at one with blood. Valves suggested the action of pumps, and Harvey confessed that he was led to the idea of the heart as a pump when he held the heart's valves as integral components of the heart's structure and function. The idea of a circular flow of the *same* blood was clinched by quantitative reasoning, which he used with telling effectiveness. In fact, his European supporters Andrea Argoli and Jean Martet made clear that the only reason for their endorsement was "the argument from quantity."

From the outset, Harvey disavowed the essential elements of Galen's physiology. There was no Galenic fire in the heart and therefore there was no need for fuel or fermentation. Heart and arteries had no respiratory function. There were no pores, neither in the ventricular septum nor in the skin. Since both ventricles contracted and expanded together, there was at no time a pressure gradient to drive blood through the septum. He also discounted Aristotle's third ventricle and the traditional argument that animals could have undergone such great alterations over the passage of time as to lose an entire heart chamber.

HARVEY BEGAN WITH VIVISECTIONS ON SNAKES, IN WHOM THE GREAT vessels were conveniently arranged for observation.[4] When one was laid open and the aorta tied in the middle with a strong knot or pinched with forceps, blood piled up before the site of the ligature on the side of the heart, and the organ itself became filled to bursting. At the same time, the aorta below the ligature became flattened and appeared empty. It was obvious that not only the distal aorta but also all the veins, which appeared flattened as well, were not receiving blood. When the ligature was untied or forceps removed, the swelling of the heart disappeared and the aorta filled up. When the same experiment was performed with the tie around the "main vein" (the inferior vena cava), the opposite effects took place. The vein below the ligature swelled up, and the section above the ligature going *toward* the heart flattened. The heart itself, receiving no blood, became flat as did the aorta. When the ligature was removed, the color and size of the heart and the great vessels were promptly restored.

In a series of now-famous ligature experiments, illustrated in his book, Harvey showed that the part of a human arm distal to a tourniquet always became swollen with blood when the superficial veins of the arm were compressed, but the arm paled and appeared emptied of blood when the more deeply located arterial flow was obstructed. That was a vital observation. It disclosed that the valves in the veins directed blood flow only in one direction, *toward* the heart. Blood always filled an empty vein from the periphery. In the *external* jugular vein in the neck, which has valves, they permitted a one-way flow only *toward* the heart.

If valves in the veins permitted blood flow only to the heart, and the valves in the heart allowed blood to pass only into the arteries, then there could only be a one-way movement of blood—*from the veins through the heart to the arteries*. Hence, the *same* blood just had to return from veins to arteries, completing a literal circulation. It was unlikely that blood could be continuously prepared in the heart chambers as Aristotle had taught, or in the liver as Galen had believed, and be continuously used up at the ends of the arteries.

But if the muscular heart pumped blood into the arterial system, how was blood returned in the veins? Harvey correctly deduced that venous return was promoted by the contractions of the leg muscles, which provided a force that no longer proceeded from the heart:

> Blood is expressed by the movements of the limbs and the compression exerted by the muscles, from capillary veins into venules, and thence into comparatively large veins, and it is thus more disposed and prone to move centrally than the opposite.[5]

His experiments and conclusions made it theoretically necessary that there be peripheral pathways for blood flow from the smallest visi-

ble arteries to the smallest visible veins. But Harvey could not demonstrate them, and critics would hold that against him: there was no reason
to believe the circulation model until Harvey could demonstrate some
form of direct peripheral vascular continuity.

CIRCLES, CROWNED BY THE SUPREME AUTHORITY OF ARISTOTLE, HAD
been applied for centuries to an understanding of the nature of the
universe. According to Aristotle, all celestial motion was circular. On
the other hand, he had maintained that natural terrestrial motions
possessed a beginning and an end.

A symbolic expression of circular movements was published by the
philosopher Giordano Bruno, whose spectacular trial and fiery death at
the stake in Rome could hardly have escaped any academic at the time,
including medical students like Harvey at Padua. Like Servetus, Bruno
denied the Trinity, the divinity of Christ, and the virginity of Mary, as
well as predetermination and transubstantiation. He described a circle as
the first principle and root of all geometric figures because it expressed,
at the same time, a whole and a part, a beginning and an end, a central
point and a circumference. In a lengthy dissertation on air and "spirit,"
Bruno proposed that the spiritual life force effused from the heart as
from the center to the periphery and then returned from the periphery
to the same center, following the pattern of a circle. And for all that, he
was tried as a heretic and executed.

Harvey linked his discovery to the cosmic theories of Aristotle.[6] The
motion of blood, he held, could be called circular in the same way that
Aristotle (and Ptolemy) had conceived the motion of planets, and air
and rain, too, followed the circular motion of the stars. From the importance he gave to that Aristotelian analogy, it is apparent that Harvey
found in it both confirmation and comfort. At the same time, as the
idea of a circulation was steadfastly taking form, he searched strenuously for examples of circular motion in terrestrial beings in order to
give them parity of status with the supposedly superior heavenly bodies.
The succession of individuals constituting a species was yet another

form of circular motion that emulated the movements of the heavenly bodies: "From frail and perishing bodies an immortal species is engendered," he reiterated in his later work *On Generation*. Changes in state, as from liquid to gas, were also circular, since the space created by the more widely dispersed gas particles permitted other gases or fluids to flow into those spaces: new substances succeeded departing substances, thereby preventing the formation of a vacuum. Franciscus Piccolomini of Padua, a philosopher who came from a learned family, had also written that the motion of animals resembled and retained the pattern of circular motion.

ON THOUGHTS REGARDING MATTER AND LIFE, HARVEY PROPOSED that blood, not the heart, was the primary cause and physical vector of life, the first to live and the last to die, and the primary seat of the soul. In the chick egg, he observed, the heart appeared only *after* the blood was created, a direct contradiction to Aristotle. The heart was a reservoir for heat, but blood was the substance in which heat first and most "abounded and flourished," from which all other organs and tissues were "cherished" and obtained their life.

As for pneuma, Harvey saw no need to consider it as anything distinct from blood. In his mind, blood and pneuma were one and the innate heat *was* blood. Criticizing Galen, he confirmed that blood could not pass through the septum:

> But, in faith, no such pores can be demonstrated, neither, in fact, do any such exist. For the septum of the heart is of a denser and more compact structure than any portion of the body except the bones and sinews.[7]

Both ventricles contracted and expanded *together*, thus precluding a pressure gradient between them, and because the septum had its own system of arteries and veins, it was unnecessary for blood to pass through it. Finally, Harvey cut away the wall of the contracting left ventricle from the heart of a living dog to expose the septum and showed that no blood came through the septum from the beating right ventricle.

HARVEY PREFERRED TO EXPERIMENT ON COLD-BLOODED ANIMALS because their hearts beat slowly and continued to beat even after they were removed from the bodies, making it significantly easier to study the details of their motion. Having patiently observed hearts for many hours at a time, he discerned by sight and touch, as Colombo had done, that, whereas *diastole* (relaxation) had hitherto been regarded as the essential movement of the heart, that role was properly assigned to *systole* (contraction), a phenomenon best observed in dying hearts, where, at the point of death, they clearly twisted upward and to the left during contraction. Dying hearts had an unequaled advantage of allowing movements to be studied in slow motion, and Harvey observed that the organ exhibited four motions: the two auricles moved simultaneously and then the two ventricles. When the tip of the ventricle was cut off, each auricular contraction was *followed* by a corresponding spurt of blood from the ventricle. Blood was *driven* into the ventricle by auricular contraction, not drawn into it by the faculty of attraction that Galen had invented.

In the second chapter of *De motu cordis*, Harvey summarized his observations and concluded that the heart moved like a muscle and its essential function was to propel blood. Even when the excised heart of an animal was held in the hand, it was felt to harden when it contracted, just as a muscle did. In eels, too, who had only a single ventricle, Harvey saw and felt the still-beating extracted heart harden rhythmically for a few instants and contract like a muscle, and, with each contraction, it became whiter as it ejected the blood. That was the mechanical aspect of Harvey's revolution in the life sciences, which had notably begun with Leonardo: the notion of a motive force proceeding from the center with effects at a distance throughout the entire body, rather than invoking local attracting forces.[8] That was directly contrary to Galen, who had held that it was the vital spirit in blood that caused the blood's motion, with the heart serving mainly as the organ that prepared the vital spirits and responded passively to the movement (expansion) they caused.

And when, at last, a heart had become completely still, it seemed to Harvey that he had just seen a beautiful machine come to a halt. We move in a world that deserves our constant wonder. Our very life is an eternally renewed miracle.

Aristotle had written that all vessels were subordinate to the soul in the heart; the heart commanded and the arteries obeyed. Harvey argued that vital motion had nothing to do with the soul or the brain. Rather, it was autonomous and intrinsic to the heart muscle itself, sensitive and itself alive, whose activity could only be *modified* by the brain and nerves. Within that context, Harvey designated the *arché*, the single source or origin, of the vital motion to be the heart itself, which he held to be only a pump. The *arché* was not, as Aristotle taught, in the self-moving blood being the "life-stuff" of the organism.[9]

Heart motion was unconnected from the movement of the lungs, whose function was respiration. In fish, which had only one ventricle and no lungs, blood passed directly from the veins into the heart and from the heart into the arteries through the pumping action of the heart, as should be obvious to anyone who had dissected a live fish. When nature had wished to send blood through the lungs for respiration, it added the right ventricle for that very purpose.

NEXT, HARVEY DISTINGUISHED THE MOVEMENT OF THE ARTERIES from that of the heart by studying, once again, the sluggish, beating fish hearts. He saw waves of blood passing at regular intervals through the aorta, which became transiently dilated and, after the passage of each wave, resumed its original shape. Contraction of the heart seemed simultaneous with the expansion of arteries and the arterial pulse. It therefore seemed highly probable that it was the heart contracting *like a pump* that caused arterial expansion, an explanation radically different from the Galenic system in which arteries were held to expand and suck blood into them like a bellows.

Harvey confirmed that in a cut artery, a spurt could be seen at the precise moment of ventricular contraction and not during that chamber's expansion. He noticed a pause in the arterial pulse when the ventricles themselves ceased to contract and that the magnitude of the pulse could be altered by manual compression of the heart. The variability of the pulse depended on the variability of such compressions. Pulsations of the pulmonary artery were likewise related to the contractions of the right ventricle and stopped when the right ventricle ceased to pulsate.

Harvey was the first to observe that, in a dead pigeon, the right auricle kept pulsating for a whole hour after the ventricle had ceased. He also benefited from observing a beating heart within the body of Viscount Montgomery's eldest son who, as a child, had suffered a severe fall that had broken the ribs on his left side, leaving a permanent gaping wound. Harvey brought the lad to court so that King Charles might watch the heart's movement and even touch the living ventricles with his royal finger!

A PHYSICAL CIRCULATION OF BLOOD WAS NOT SOMETHING THAT WAS directly visible. It was conceptual: a circle was a symbol that ended at the point at which it began, and Harvey realized that he had to make his argument for a circulation cogent. It was an instance when it was necessary to make an immense leap between observed facts and the secrets they concealed. It was here that he hopped off the Galenic bandwagon and advanced into territory few in the biological sciences had claimed before, by thinking in terms of quantification. Galileo was teaching mathematics at Padua during Harvey's student days, and it is highly likely that Harvey was exposed to at least some of that scientist's quantitative principles and methods. Galileo's demonstrations, too, emphasized that explanations of motion were found in the mathematical structure within it, and experiments, even if they were "thought experiments," were necessary to tease out the implicit mathematics.

If the system of arteries and veins was analyzed hydraulically by thinking of the heart as a pump, the vessels as pipes, the heart valves and the valves in the veins as mechanical valves, and blood itself as simply a fluid that was pumped, then quantitative deductions on blood flow through that closed system could be made. Hitherto, Harvey's argument had been based virtually on circumstantial evidence—the fitting of every structural key into a functional purposive lock, backed by a wealth of comparative anatomical observations and ligature experiments. But the key that could unlock the last door to the secret of a circulation must be a *quantitative* assessment, which would demonstrate the *quantitative* necessity for a completely closed circuit. According to Aubrey, Harvey was well versed in mathematics and had made himself master of William Oughred's widely used textbook *Clavis math* on algebra and arithmetic. Even in his old age, Aubrey wrote, he perused it and worked on its problems, and the volume was always in his "meditating apartment." From quantitative arguments that blood *must* circulate, Harvey demonstrated that blood *does* circulate.

It was not the first time that human physiology was quantified. Nicholas of Cusa had described the first biological application of the physical balance in 1450, and his contemporary at Padua, Santorio, too, had appealed to measurement. So had Van Helmont. Given those precedents, Harvey began by measuring the whole amount of blood in the animal body. He simply cut a prominent vein and collected all the blood that drained out. The amount was finite; indeed, by severing any artery, all blood could be depleted within half an hour. Thus, blood could not be continually formed by the liver as Galen had taught. By observation and experiment, Harvey demonstrated that during every cycle of expansion and contraction, the heart received and expelled a measurable quantity of blood.

In a magnificent thought experiment that he outlined in chapter 9 of *De motu cordis*, he reasoned that, at each contraction, let's suppose the left ventricle pumped out only 2 ounces, because, when Harvey emptied the human left ventricle of all blood at dissection, he collected about 2 ounces.[10] From the number of heartbeats per minute (about 72) and per hour (60 × 72), he estimated that the heart ejected 2 × 72 × 60 ounces = 8,640 ounces = 540 pounds every hour! That was more than three times the weight of a heavy man. And that amount was perpetually repeated as the heart contracted like clockwork. That cumulative quantity was much larger than all the blood present in the entire body when drained! In fact, the arteries would be ruptured (the absolute numbers did not matter; Harvey's point was that the numbers added were very large).

He wrote:

> But let it be said that this does not take place in half an hour, but in an hour, or even in a day. Any way it is manifest that more blood passes through the heart in consequence of its action, than can either be supplied by the whole of the ingesta or can be contained in the veins at the same moment.[11]

Such a stupendous quantity could not possibly be continuously formed in the liver from ingested food within that time as Galen had taught. No one could consume 540 pounds of food and drink in one hour! Nor could that amount be absorbed at that rate by the tissues as in Aristotle's analogy of an irrigation system. The only way out of such an absurdity was to infer that, contrary to Aristotle's and Galen's beliefs that blood was consumed at the periphery and was instantly renewed, the *same* blood must return to the heart (through the veins) and be pumped out again (through the arteries). Blood must *circulate* continuously. The magic of the circle seemed the only true secret. Neither observation nor investigation was the decisive step—the thought experiment sufficed, although Harvey commented that he had made many observations on the amount of blood ejected during a single beat and the factors involved in increasing or diminishing that volume. Harvey made his own estimates intentionally low so that he could show without doubt the vast amount of blood that Galen's theory would require the liver to produce and the tissues to take up.

Here was an entirely new viewing of facts. Ancient theories had been qualitative, not dynamic in character, and had been conceived without regard to movement and time. Harvey extended the new era of Galileo to the life sciences, whereby functional thinking was brought about by the modern employment of measuring, weighing, and computation. To that extent, Harvey was a "modern," a man dedicated to the solution of a *problem of purpose*.

THE HEART WAS A MASTERWORK, AN EMPEROR OF ALL ORGANS, WITH powerful muscles that, of their own accord, propelled blood to the farthest extremities of the body, even in elephants and whales. The valves in the veins and in the heart allowed blood to flow in one direction in much the same way as "the two clacks [valves] of a water bellows" allowed water to be raised in one direction. From his meticulous measurements of the sizes of blood vessels, the timing of the pulse, and the quantities pumped from the ventricles and auricles in different species, it was inevitable that arteries and veins with the heart all formed part of a single system!

Harvey remembered—indeed, nothing could tarnish—the exact moment when the truth had struck him, and his exhilaration; the same emotion derived when the cold, logical mind sees the beauty of a geometric theorem or a mathematical equation. He wrote:

> I frequently and seriously bethought me, and long revolved in my mind, what might be the quantity of blood which was transmitted, in how short a time its passage might be affected, and the like; and not finding it possible that this could be supplied by the juices of the ingested aliment without the veins on the one hand becoming drained, and the arteries on the other hand becoming ruptured through the excessive charge of blood, unless the blood should somehow find its way from the arteries into the veins, and so return to the right side of

the heart: I began to think whether there might not be a MOTION, AS IT WERE, IN A CIRCLE.[12]

The inference was inevitable: (1) a fluid flowing constantly through a point (namely, the heart) will either (a) be continually generated or (b) move in a circle. (2) Harvey demonstrated that (a) cannot be true; (3) hence, the fluid must move in a circle.[13] The circulation had to be demonstrated from premises in order to be counted as certain knowledge; the mere agreement with observation was not enough.

It was now clear to Harvey why the heart was emptied when the vena cava had been ligated, why it was filled to distention when the aorta had been tied, why a middling ligature that compressed only the veins and not the arteries made a limb swell turgid whereas a tight ligature that blocked the arteries made it bloodless and pale, and why the whole blood in a body could be drained by a single opening in a single blood vessel. It was now clear what the purpose was of those valves in the veins that old Fabricius had demonstrated at Padua, but whose function he had not correctly understood.

HARVEY ALLOWED ALMOST A DECADE TO ELAPSE BEFORE HE SUBMIT-ted his manuscript for publication to Wilhelm Fitzer of Frankfurt-on-Main. The choice of Frankfurt, most likely suggested by his friend Robert Fludd, was determined by its being one of the centers of the continental book trade and the scene of an annual book fair, as it still is today. Harveyan scholar Gweneth Whitteridge, however, holds that the book was published at Leiden.[14]

Every day before he sent it, and for some days thereafter, he awakened with a strange feeling, a vague anxiety, such as a novice actor might experience before going onstage. The road for the manuscript was not the smoothest possible. Having survived the fogs and storms of the English Channel, it had to find its way by lumbering cart through the Dutch theater of war, and then the printer had to decipher the handwritten text without being able to discuss the illegible passages with its author. The *Exercitatio anatomica de motu cordis et sanguinius in animalibus* (*Anatomical exercise on the motion of the heart and blood in animals*) was a "new and unheard opinion concerning the motion of the heart and the circulation of the blood," as Harvey explained in the preface.[15]

"I cannot sufficiently wonder how it could have happened that Galen, a man systematic more than can be expressed, and an exceedingly keen observer of nature, could have fallen into such serious errors," Colombo reminisced in *De re anatomica*. The greater wonder is that it took two thousand years for experienced natural philosophers, capable of painstaking accuracy in both perception and interpretation, to make what should have been superficially a straightforward case of observation and deduction. No sophisticated technology or complex mathematics was required; the facts were staring them in the face. It took even Harvey many years of work to make his discovery after he left Fabricius.

twenty-two

Closing the Ring

*Well, my son, how do you think that Harvey was
regarded by his contemporaries? — As a dissector
of fleas, snakes, butterflies and insects!*

—JULIEN OFFRAY DE LA METTRIE

HARVEY WAS FIFTY YEARS OLD, SOMEWHAT OLDER THAN MANY OF his peers, when he published his discovery. When he began lecturing at the College of Physicians in April 1616, contemporary ideas about blood motion still rested on the Homeric ebb and flow derived from Galen, who had, over centuries, provided what seemed to be a complete and plausible explanation for the assimilation of food, the creation of blood, the distribution of nourishment, the generation of the heartbeat and the pulse, and the production and distribution of heat. A significant modification was the pulmonary transit of Colombo, and Caesalpinus had even coined the word *circulatio* to describe it. Harvey was aware of those observations. He may have learned about them at Padua, where Colombo had succeeded Vesalius, who, in turn,

171

had been succeeded by Fallopio and then by Fabricius, Harvey's own teacher. He had learned of the valves in the veins from Fabricius himself, although their function had been totally misjudged.

Aware of that vast legacy, Harvey began his own investigations, and what progress he had made by 1616 he recorded in the notes for his lectures. None of the arguments expressed were entirely new. His real contribution lay in his approach, which was founded on vivisection. He was so obsessed by "ocular demonstration" that not a week passed without a human or animal corpse being laid out on one of the large wooden tables in a room at the back of his house. It was rumored that he had even performed autopsies on his own father, sister, and "a close friend." He was impressed by Colombo's correct observation of systole as an active contraction, with diastole being a passive dilation, and he had recognized the proper relationships between the pulsation of the heart and the arterial pulse.

Now he began, with minute care, his own studies on the pulsation of ventricles and arteries in frogs and fish. By the time of his Lumleian Lectures, his observations were virtually complete. There was no discussion of the auricles in his lecture. There was only a single experiment using a ligature. The valves in the veins he mentioned once in his notes with respect to the veins of the leg, but Harvey ascribed to them their conventional function as proposed by Fabricius. He accepted the pulmonary transit of blood and gave credit to Colombo, and it was likely that, by 1616, Harvey was already convinced about the absence of the septal porosities. Although Vesalius had doubted their presence and Colombo had altogether dismissed them, most of Harvey's contemporaries still believed in them.

All in all, when Harvey wrote the Lumleian notes, it is doubtful that he had any firm notion of a circulation. Indeed, he despaired whether he would ever be able to understand even the heart:

> When I first gave my mind to vivisections as a means of discovering the motions and uses of the heart, and sought to discover these from actual inspection, and not from the writings of others, I found the

task so truly arduous, so full of difficulties, that I was almost tempted to think, with Fracastorius, that the motion of the heart was only to be comprehended by God.[1]

It is worth mentioning here of a controversy that would be generated by a statement written in his own hand in his personal copy of the published lectures: what is now referred to as *"f.80 v."* It is inscribed on an otherwise blank *verso* page, and its contents are not reflected in any of the printed text. The inscription mentions the passage of blood from the arteries to the veins, the comparison of the heart action to a water pump, and the perpetual motion of blood in a circle. If all those facts were indeed present in Harvey's mind at the time of the lectures, then it must be concluded that he had realized the circulation at that early date and therefore would have mentioned it. It is now almost certain that the handwritten statement is an addition that Harvey wrote later sometime *after* the lectures, perhaps between 1616 and 1619, when Harvey began thinking of a circulation.

IF, AS HAS BEEN SUGGESTED, THE HYPOTHESIS MATURED BETWEEN 1619 and 1625, then Harvey was between the ages of forty-one and forty-seven. During those years, the issue of visible peripheral connections, or anastomoses, between veins and arteries to "complete" the circulation had continued to defy and vex him. His experiments and conclusions made them a reasoned necessity, but he had come to accept that final and complete proofs in science may never arrive. Sometimes, the best one could do was to construct a model that accounted for all the available evidence and then test the model repeatedly in novel ways, thus moving closer to its acceptance.

More than a millennium earlier at Alexandria, Erasistratus had declared that the peripheral veins and arteries joined up. Galen, too, had spoken of peripheral *osculations*. More recently, Andreas Caesalpinus, in his usual cavalier fashion, had written about blood vessels that "did not end but rather carried on." Without a powerful microscope, Harvey could not see what connected arteries and veins. But he speculated cautiously, as had Galen, that either there were connections between the vessels (small arteriovenous *inosculations* or *anastomoses*), or else there had to be porosities in the flesh (*porositatis carnis*; *caecas porositates*) into which blood seeped from the arteries into the veins. In his letter to the French anatomist Jean Riolan, he endorsed the latter view because, despite boiling the liver, spleen, lungs, and kidneys to render the tissues maximally friable, he had failed to detect any visible peripheral connections. Although Galileo's microscope was already afoot since 1610, Harvey did not live long enough to learn of Malpighi's discovery of those very anastomoses he had hoped for.

Born into an affluent Protestant farming family in 1628, the same year as the publication of *De motu cordis*, and orphaned before he would commence his medical studies, Marcello Malpighi graduated in medicine and philosophy from Bologna, where he was introduced to Harvey's new physiology.[2] He had no sooner undertaken his duties as public lecturer in anatomy than Ferdinand, Grand Duke of Tuscany, created a chair for him in "Theoretical Medicine" (physiology) at Pisa. Microscope and bundles of notes under his arm and his new bride by his side, the twenty-eight-year-old scholar journeyed by sea to a new university, not yet famous, but which promised acquaintance with new learning and technologies, such as those proposed by the irascible Giovanni Alfonso Borelli of Naples, a student of Galileo, who was already professor of mathematics at Pisa. It would turn out to be an exhilarating experience!

Borelli, born at Naples in 1607, had studied at Pisa and later taught at Messina, Pisa, and Florence, where he became an active member of the Accademia. Returning to Sicily, he was politically active against the Spanish occupation and had to flee to Rome, where, after a brief period under the protection of the Catholic queen Christina, he died in a religious house in 1679. Twenty years senior, Borelli became a great influence on Malpighi.[3] Together, they vivisected living animals in his parlor and observed the "subtle movements of the organs." But the humid air of the yet-undrained Pisan swamps proved too oppressive for the Bolognese pupil. Modest to the point of timidity, Malpighi returned quietly home. A double convex lens, and later a pair of them, allowed him to see blood passing from the arteries to the veins, so making Harvey's *sanguinis circuitum* complete.

HARVEY HAD USED SOME KIND OF MAGNIFYING LENS IN HIS STUDY OF the heart: "Nay, even in wasps, hornets, and flies I have, with aid of magnifying glasses, myself seen, and made many others see, the heart pulsating at the upper part of what is called the tail," he wrote. Robert Hooke, Boyle's and Newton's assistant, too, made valuable microscopic studies of biological phenomena. In 1661, Malpighi applied his lenses to a study of those internal organs that were transparent. He examined lower organisms in the belief that they would shed light on the nature of higher animals. He was the first to inject water in the smaller arteries to wash out the blood and so make the tiniest vessels more visible. The lining membrane of the frog's abdominal cavity, called the mesentery, when spread out under his instrument showed a network of unsuspected tiny blood vessels. The frog's lung was not much different. Focusing his lenses, he saw vessels joined together in a ringlike fashion that proceeded from the artery on one side to the vein on the other so that they "no longer maintained a straight direction and appeared as a network made up of the continuations of the two vessels."

Malpighi announced his findings modestly in two epistles (letters) dedicated and addressed to his former mentor, Borelli, who had urged their publication. His style was neither lucid nor dramatic, nor did he appear too excited over his discovery, perhaps because it was long taken for granted that such anastomoses must exist. The first, a wordy and meandering epistle titled *De pulmonibus* (*On the lungs*), came out at Bologna in 1661, more than thirty years after Harvey's book and nine years after his death. In it, he announced "a few little observations that might increase the things found out about the lungs."

Those few little observations were the detection of two separate interconnecting vessels and vesicles: the air and blood networks of the lung, and the movement of blood in the capillary network in the lungs of a living frog while the heart was still beating. He completed his studies on lungs that had been dried after the pulmonary veins were tied so that the vessels were full of blood. The membranous, or pleural, surface of the dried lung where a single alveolus jutted out showed a "marvelous" network, entwined and connected, when observed under transmitted light. Malpighi blew air into the lungs and noted the pulmonary vessels standing out like a sculpture in relief, like the branches of a tree. He injected mercury and colored fluids into the pulmonary artery and saw the colors enter the veins and ooze through the investing membrane of the capillaries into the interstices of the lung. However, he still had some reservations in deciding whether the finest vessels anastomosed, or "gaped into the substance of the lung."[4] He inferred that the lungs were, by nature, a storehouse of blood, made not for cooling as had been held since the time of the ancients but rather for its proper mixing.

In the second epistle, his doubts were resolved when, viewed "by use of a more perfect lens," he was sure of the union of blood streams in the lungs.

He wrote:

> The smallest vessels . . . mingled annularly. . . . Here it was clear to sense that the blood flows away through the tortuous vessels, that it is not poured into the spaces but always works through tubules, and is dispersed by the multiplex winding of the vessels . . . the same holds true in the intestines and other parts: nay, what seems more wonderful, she [nature] joins the upper and lower ends of the veins [vessels] to one another by visible anastomosis.[5]

The same observations were confirmed in a frog's urinary bladder. Malpighi felt that he now had enough evidence to draw general conclusions "for if Nature turns the blood about in vessels and combines the ends of the vessels in a network, it is likely that in other cases an anastomosis joins them." He concluded, "I can better make use of that [saying] of Homer for the present matter—'I see with my eyes a work trusty and great.'"

All the passages required for a circulation of blood were now described. The blood vessels and their contents were a single system around which natural philosophers could wrap their minds concerning the mysteries of nutrition and warmth and the provisioning of the body as a whole.

Malpighi corresponded with the Royal Society in London whose secretary, Oldenburg, wrote to him: "Our company of philosophers thinks that you are treading the real paths leading to a true knowledge of secrets. . . . You devote your mind and hands to observing accurately and eviscerating minutely the things themselves."

MALPIGHI WAS NOT THE FIRST TO DISCERN THE CAPILLARIES. THE English philosopher John Locke, who was a well-respected physician and chemist, had mentioned them in the late 1650s in his book *Circulatio sanguinis*. "Take a frog," he wrote, "[and] strip it you may see ye circulation of bloud if you hold him up agt ye sun." But Malpighi was the first to remark upon them in a systematic way. In a later tract, Malpighi described small, flat red cells within the blood vessels in the mesentery of a hedgehog. They appeared as globules of fat of definite outline, reddish in color, presenting a likeness to a "chaplet of red coral." But his lenses were not powerful enough to do more than characterize blood as a liquid consisting "of an almost infinite number of particles."

Although he was the first to observe the red blood corpuscles, he mistook them for fat globules. In that discovery, he was anticipated by a Dutchman, Johannes Swammerdam of Amsterdam, whose *Biblia naturae* (*Bible of nature*) published posthumously in 1738 contained a record, entered in 1658, on the blood of a frog where he had perceived "the serum in which floated an immense number of rounded particles, possessing the shape of as it were a flat oval but nevertheless wholly regular."[6] When looked at sideways, they resembled transparent rods. He died young in 1680 and published little during his lifetime.

It would remain for another Dutchman, "Father of Microbiology" Antonie van Leeuwenhoek, to be the first to see "corpuscles" flowing in the capillary stream.[7] Leeuwenhoek reveled in the joy of microscopy: "A sight presented itself more delightful than any mine eyes have ever beheld," he wrote. A self-taught, humble draper in Delft, he devoted his free time to building simple single-lensed microscopes of remarkable precision and resolution but that, nonetheless, offered only imperfect definition and were not at all achromatic. It was only when he extended his research to the "tail of frog worms" (tadpoles) and then to fish, and was forced to devise an aquatic microscope which allowed the dissected parts to separate more readily under water, that he saw more than was possible by the lens alone, and "it plainly appeared to me, that the blood vessels I now saw in this animal, and which bear the names of arteries and veins, are, in fact, one and the same continued blood vessels."[8] Both the narrowness and the profusion of those microscopic pathways caught his attention. Whenever possible, he applied measurements and calculations to his observations and tried to determine the volumes of blood and the speed of blood streams: in an eel that was eleven inches long, he estimated (erroneously) that blood circulated around its body a little over thirteen times every hour!

Leeuwenhoek reported his "trifling observations" in letters written in Dutch to the Royal Society, beginning in the summer of 1673. Ignorance of Latin no longer excluded an inventive draper from the community of scientists. He had seen the single corpuscles follow each other, compressed and in single file, through the narrowest vascular channels, and he was the first to recognize that they were responsible for the redness of blood. He concluded that it must be the red corpuscles that delivered the "more subtile juices" in blood to the tissues through the capillary walls for their nourishment. Blood, returning in the veins and deprived of those juices, appeared blackish. That was a vital observation—it was a substance in the corpuscles that was, in some way, responsible for the color changes between the arterial and venous blood.

Despite their trivial appearance, capillary function would prove to be multiple, challenging, confusing, and highly capricious. The variability of capillary blood flow, the question of independent capillary contractility as well as the capillary blood pressure, the passage of substances through capillary walls, and, finally, the nature of the capillary wall as a semipermeable membrane would culminate in Nobel-winning research two centuries later.

twenty-three

Ahead of the Curve

The beginning of truth is to wonder at things, as Plato says in the Theatus, as Matthias in the Traditions, advising us: Wonder at things that are before thee, making this the first step to further knowledge.

—CLEMENT OF ALEXANDRIA

TO HARVEY'S STUDENTS AND FOLLOWERS AT OXFORD, THE IDEA of a circulation opened interesting avenues.[1] Indeed, the Oxford Chemists, as Boyle, Hooke, Lower, and Mayow were called, endorsed and extended the Harveyan discovery soon after its publication in their own research. The millennia-old practice of bloodletting at a specific site to cure a local lesion was now nonsensical because blood circulated through the entire body. On the other hand, the swift action of poisons, or the rapid dissemination of infection from a septic wound, could be logically explained. In fact, a drop of intravenous poison should be "carried by the circulated blood to the Heart and the Head" faster than the same amount taken by mouth. It followed that medications, too, could be more effective if introduced intravenously. By 1651, two decades after Harvey's publication, that novel idea had suggested a new therapeutic technique—that of blood transfusion. It led to attempts at blood volume replenishment after heavy hemorrhage.

Since antiquity, blood was revered as a semi-magical vital fluid of great power. Philosophers had extolled the potential rejuvenating effects of "putting new blood into old veins." Since it was held that bad blood caused debility or chronic disease, bloodletting to release noxious agents had seemed a sensible approach to effect a cure. It now seemed even more logical that if the blood of a diseased person were *replaced* from a healthy animal, improvement in health could ensue. As early as 1615, before Harvey's publication, a German physician at Halle, Andreas Libavius, had applied silver tubes to bleed a healthy young man into the veins of an old and infirm patient. Closer to home, the English experimentalist Francis Potter had conducted a series of varyingly successful experiments at his rectory in Wiltshire. He collected blood from an animal in a small bladder and then transfused it into another animal using a variety of ivory tubes and quills.

By 1656, Robert Boyle and Christopher Wren, the chemist and the architect, were already discussing intravenous injections at Boyle's Oxford lodgings. Sometime that year, a year before Harvey's death, Wren and his colleagues inserted a hollow, slender quill, attached to a small animal bladder, into a vein in a dog's hind leg. They infused a warm solution of opium in ale, thus stupefying the animal, which eventually recovered. The inevitable next step "to open veines and spout medicins into it" followed. With his fellow Oxonian Timothy Clarke, who had moved to London, Wren injected "different kinds of waters, beers, milk, whey, broths, alcohols, wines, and even blood itself." Antimony sulfate (*Crocus metallorum*) was an emetic and, predictably, a dog injected with it "immediately fell a Vomitting, & so vomitted till he died." That event was archived in the *Proceedings of the Royal Society*.

The Cornishman Richard Lower, another enthusiast, even wondered whether a dog could be kept alive "without meat, by syringing into a vein a due quality of good broth," a suggestion that kept intriguing Boyle as did the potential of transmutation in an animal's appearance and behavior following repeated transfusions between species. Such questions of hybrid species also struck the diarist Samuel Pepys, who mused over the outcome if "the blood of a Quaker [was] let into an Archbishop?" In the autumn of 1657, Clarke and Wren convinced a maidservant—"an inferior Domestick that deserv'd to have been hang'd"—at the residence of the French ambassador, the Duke of Bordeaux, to submit to a covert intravenous infusion of *Crocus metallorum*. But she fainted prematurely, and the experiment was hastily aborted. Wren would cease all future transfusion experiments thereafter.

IN FEBRUARY 1666, RICHARD LOWER, THE "HERO OF TRANSFUSION," publicly revived, with a blood transfusion, a dog that had been bled almost to the point of death. That was the first successful dog-to-dog public transfusion conducted at Oxford. He had spearheaded the movement in 1657, the year of Harvey's death, but without success for technical reasons. Lower's demonstration was repeated before a select audience in November by another *transfuseur*, Dr. Edmund King, along with Robert Hooke, the society's curator, and Samuel Pepys recorded it. The event was followed by the first interspecies transfusion between a dog and a sheep.

Beginning 1667, the Royal Society was fully committed to blood transfusions between species. Although Lower normally prescribed transfusion from an artery of a donor into a vein of the recipient, King improvised by transfusing from vein to vein. Techniques of transfusion were summarized in the fourth chapter of Lower's textbook *De corde* (*On the heart*), in which he claimed full credit for the priority of this research. On November 15, 1666, Oldenburg, the society's secretary, announced, "We have now conquered ye difficulty of bleeding one animal into another."

It wasn't long before Harvey's friend George Ent proposed that the society try to experiment on "some mad person in the [lunatic] hospital of Bethlam." On November 23, 1677, Lower gave a breathtaking demonstration before a select invited audience of forty witnesses including physicians, members of Parliament, and a bishop, who gathered at the society's quarters at Arundel House in the Strand. He transfused blood from a young sheep into "a poor and debauched" divinity graduate named Arthur Coga, who had been a student at Pembroke College, Cambridge. Being educated, it was felt that he could give a reliable account of his experience, though Pepys did remark that "he is cracked a little in the head, though he speaks very reasonably and very well." Perhaps a transfusion could help cure his mental illness by replacing his overheated blood by another creature's cooler one.

The heavy-drinking Coga survived the celebrated event without consequence, and Boyle was informed in a letter that he "took a cup or two of sack before, and a glas of wormwood wine and a pipe of Tobacco after ye operation." The following day, Coga reported "that ye same day of this operation he had 3. or 4. Stools" as proof of his full recovery. Coga, who received a fee of one guinea for his part, quipped that he was receiving "ye bloud of ye Lamb"—symbolically, the blood of Christ— and agreed to a repeat demonstration at a higher price before a larger "strange crowd both of Forrainers and domesticks" that was recorded by the diarist John Evelyn. A somewhat less successful event was entered by Pepys, who, in the company of the grand old Duke of York, witnessed an experiment of killing a dog by letting opium into his hind leg. Pepys recorded that "Mr. Pierce the surgeon and Dr Clerke [Clarke] did fail mightily in hitting the vein, and in effect did not do the business after many trials; but with the little they got in, the dog did presently fall asleep and so lay till we cut him up."

BOLOGNA, TOO, GOT CAUGHT UP IN THE ENTHUSIASM. IN EARLY 1667, the natural philosopher Giovanni Cassini performed a successful

transfusion between sheep, and, by that summer, interspecies experiments were well underway. A surgeon named Griffoni successfully
transfused his own deaf pet spaniel, who was failing to thrive, with a
lamb's blood. In the fall of the same year, the respected doctor Paolo
Manfredi was invited to demonstrate a transfusion at Rome before the
niece of Cardinal Mazarin. In 1705, the Bolognese anatomist Carlos
Fracassati, a colleague of Malpighi, published his experiences of injecting various strong chemicals into the veins of dogs and observing their
effects, both at vivisection and postmortem.

In Germany, Johann Daniel Major, who had graduated from Padua
a generation after Harvey and was now professor of botany, chemistry,
and theoretical medicine at Christian-Albrechts-Universität at Kiel, began to infuse medications intravenously. In 1664, he developed a rather
modern-looking syringe using a silver piston and a cylinder that had a
cannula-like spout at one end. Blood, when drawn into the cylinder
by partially withdrawing the piston, could be injected into a recipient
through the spout by pushing the piston; the "needle" was a tiny silver
pipe. He recommended adding a grain of volatile "staghorn salt" or
"Salmiak Spirit" to prevent the blood from clotting.

In June of 1667, all Europe was abuzz when word came from
France that blood had been transfused from a young sheep into the
veins of a sick fifteen-year-old boy, bringing about the latter's complete
recovery. That report was, in fact, the first fully authenticated human
blood transfusion conducted by Jean-Baptiste Denis, physician to the
Sun King, Louis XIV. But his success was short-lived. The next beneficiary, a mentally ill thirty-four-year-old man named Antoine Mauroy,
died, and Denis was accused of murder! He was saved only by royal
intervention after rumors that Antoine had been poisoned by his own
wife. But the court at Versailles, with support from the Paris Faculty of
Medicine, mandated no further human transfusions. Many more humans were transfused in Italy, but with the deaths of as many victims,
there arose a public uproar, and all further blood transfusion was abandoned for over a century.

Notwithstanding, when George Washington succumbed on December 14, 1799, to four consecutive, generous bloodlettings by his well-meaning surgeons for a sore throat and fever, Dr. William Thornton counseled that the president's corpse be revived by opening "a passage to the lungs by the trachea, and to inflate them with air, to produce artificial respiration, and to transfuse blood into him from a lamb."

Fortunately for the corpse, Washington's daughter firmly opposed the recommendation.

A WORLD ON FIRE

Science commits suicide when it adopts a creed.

—Thomas Henry Huxley

twenty-four

Pride and Prejudice

It is much easier to recognize error than to find Truth: the former lies on the surface, this is quite manageable. The latter resides in the depth, and this quest is not everyone's business.

—Johann Wolfgang von Goethe

SCIENCE HAS A WAY OF CREATING MYTHS TO EXPLAIN EXISTENCE and to address change. Galen had imagined septal perforations and an ebb and flow of blood a thousand years earlier and the world had marveled. It mattered little that what had been offered were false marvels. New ideas do not battle so much with ignorance as with presumed knowledge. Even when scientists had finally allowed themselves to be convinced that there were no such orifices, they remained firmly attached to the *idea* of a septal passage and tried to force facts into unnatural agreement with their own assumptions.

Harvey was aware of the novelty of his exposition.[1] The circulation was, in fact, a lethal blow to the whole, all-encompassing, interlocking

189

grand cathedral of theoretical physiology laid down by Aristotle and Galen and buttressed by the Arabs. But if Galen was wrong about the heart and blood, then couldn't he be wrong about all else—digestion, nutrition, reproduction, respiration? From the beginning, the new idea gave Harvey anxieties, not because it was at odds with common sense but because it was so with the common world that common sense reflects. Victorian neurosurgeon Wilfred Trotter, who made a study of the herd instinct in people, pointed out that change is "in its very essence . . . repulsive and an object of fear . . . a little self-examination tells us pretty easily how deeply rooted in the mind is the fear of the new." Freud concurred: there is an innate tendency to resist new ideas because "it contradicts many of [man's] wishes and offends some of his highly-treasured convictions. He will then hesitate, look for arguments to cast doubt on the new material, and so struggle for a while until at last, he admits to himself: This is true after all, although I find it hard to accept and it is painful to have to believe in it."

THE DIE HAD BEEN CAST. HARVEY KNEW THAT FOR ANY NEW INTEL-lectual pursuit to exert a lasting influence, it must be accepted by his peers in academic circles, who should realize the novelty of his discovery and accept its validity. Conflicts were bound to happen with Galen's supporters. Nor does the conflict confine itself to individual opponents. It goes on inside the head of every thinker and hardly anyone is totally "a modern."

Varieties of opposition to nonconformity lurked beneath the British orthodox exterior, and the men who should have helped him the most cast him aside the first. Thomas Winston, his colleague and lecturer at Gresham College, still accepted and taught about the septal pores.[2] John Collins, Regius Professor of Physic at Cambridge, ignored the new discovery. The influential Helkiah Crooke of St. Johns College, Cambridge, elected to "leave the subtle question of the passage of blood through the septal pores to philosophers." Harvey's surgical associate

Alexander Reid (Reade), the Scottish anatomy lecturer at the Hall of Barbers and Surgeons on Monkwell Street, gave a traditional account in his widely read *Manuall of the Anatomy* published in London in 1634 and reprinted, without changes, several times until 1658. One London surgeon, Sam Sambrooke, still believed in a three-chambered Aristotelian heart and even hinted that he had seen one at dissection! Small wonder, then, that knowledge of the circulation was slow to percolate when the English teachers, both in their lectures and in their popular textbooks, chose to condemn, by omission, a doctrine of which they were perfectly aware. Only the young Oxford graduate Edward Dawson elected to defend his medical diploma by adopting, as his graduation disputation, an affirmative position on the question *An circulatio sanguinis sit probabilis.*

SELECT LAYPERSONS WERE OFTEN INVITED TO THE LECTURES AND demonstrations held at the College of Physicians, and it is probable that among those who attended Harvey's lectures was the celebrated poet and preacher, the Royal Chaplain John Donne. Donne's contemporaries were greatly impressed by his wide range of learning; his poetry reflected his interest in medicine and the scientific knowledge of his day. In one of his shorter poems, *Of the Progresse of the Soule*, written in 1612, he showed an acute awareness of a problem that was vexing the most astute physicians of the age:

> Know'st thou how blood, which to the heart doth flow,
>   Doth from one ventricle to the other goe?

That this enigma was in Donne's mind just at that specific time probably indicates that some germ of Harvey's ideas had already reached him. Harvey was a firm Christian, and for the Royal Physician not to have met the Royal Chaplain would indeed have been extraordinary, especially when the chaplain was one so well read in medicine. They also had mutual friends. Donne's personal physician was Simeon Fox, who had been one of Harvey's fellow students at Padua and had witnessed Harvey's diploma.

Donne also mused "How little of a Man is the *Heart*, and yet it is all by which he *is*." Preaching at Whitehall on April 8, 1621, to a court notorious for its gluttony and heavy drinking, he used an unusual analogy to describe man's greed for material things:

> We know the receipt, the capacity of the ventricle, the stomach of man, how much it can hold: and we know the receipt of all the receptacles of blood, how much blood the body can have: so, we do of all the other conduits and cisterns of the body. . . . When I looke into the larders, and cellars, and vaults, into the vessels of our body for drink, for blood, for urine, there are pottles and gallons. when I looke into the furnaces of our spirits, the ventricles of the heart and of the braine.

When Donne mentioned "the ventricles of the heart," he was most likely referring to Harvey's many demonstrations at the college. Nobody but Harvey, in 1621, was discussing the size and capacity of the visceral organs, measuring and demonstrating "the receipt of all the receptacles of blood." Moreover, Harvey *had* called the thorax the "parlor" and the stomach the "kitchen," and *had* talked of furnaces to draw away the phlegm and raise the spirit. From Donne's sermon we can see that by 1621, the germ of a circulation had likely been installed in Harvey's mind.

Even before the publication of *De motu cordis*, the mathematician and scientist Walter Warner had published that a "perpetual circulatory pneumato-hydraulic motion" involved the regular suction of blood from the veins to the heart and its consequent dispersal from the heart to the arteries. The heart was responsible for that motion, being suited for that purpose by its muscular structure, which brought about a regular "voluntary dilatation and contraction."[3] After the publication of *De motu cordis*, a contemporary political journalist and amateur poet, John Birkenhead, versified how "immortall Harvey's searching Braine / Found the Red Spirit's Circle in each veyn." Daniel Defoe, the pioneer novelist, made the analogy between "the circulation of trade within ourselves, where all the several manufactures move in a just rotation from several countries where they are made, to the city of London, as the blood in the body to the Heart; and from thence are dispers'd again." Was Shakespeare, too, referring to Harvey's circulation when he wrote in Act I, Scene 5, of *Hamlet* of the "blood of man / That, swift as quicksilver, it courses through / The natural gates and alleys of the body"? Harvey, interested in the arts and letters and a frequent patron at Blackfriars Theatre and the Globe, may well have met the playwright through the young physician John Hall, who was married to Shakespeare's daughter Susanna. Harvey was acquainted with Ben Jonson and other poets at the Friday Street Club, which assembled at the Mermaid Tavern; Donne, Chapman, and Fletcher were invariably there.

THE FIRST MENTION OF THE CIRCULATION TO SHOW UP IN ENGLAND was his friend Robert Fludd's book *Medicina catholica* (*Universal medicine*), published in 1629 and written in baroque Latin. Fludd, in fact, preceded Harvey to the announcement of a "circulation." He was an alchemist and Hermetic philosopher (a follower of the mythic Hermes Trismegistus) who devised mechanical models to explain human pathophysiology and pursued the Hermetic motto "as it is above, so it is below."[4]

The sun, Dr. Fludd taught, traveled around the earth daily in a circle and impressed its circular motion on the winds, which were then inhaled by man. Thus, the spirit of life reached the heart and was distributed in a circular motion that imitated the divine circularity, but by no means identifiable with the Harveyan pathway. He declared, "My most dear countryman and colleague, W. Harvey, a distinguished anatomist and a profound philosopher, has confirmed and declared, with many ocular demonstrations, that blood moves in a circuit." The potential alchemical implications of the interconversion of arterial and venous blood certainly enamored Fludd, who heartily expressed his support. William Gilbert of magnetism fame, Harvey's friend and predecessor at the college, saw analogies of the organ systems of animals throughout nature and spoke of a "circulation" of humors in the viscera of the earth, whereby humors were carried to the surface in the veins (rivers), poured out through springs, and returned to the interior by gravitation.

Then the Calvinist Scot James Primrose cast the first stone. Primrose, the son and brother of physicians, had been examined by Harvey himself and passed as a licentiate of the College of Physicians as recently as December 1629. Born in France of Scottish parents, he was educated at Bordeaux and in the Paris of the anatomist Jean Riolan and had obtained a medical diploma from Montpellier. The purpose of medicine, Primrose declared, was to practice the vested learning of Hippocrates and Galen, and the new discoveries of the "circulators" (a contemptuous allusion to itinerant quacks) added nothing to that end and, indeed, destroyed the foundation of such venerable practices as bloodletting.

Writing somewhat disrespectfully, he addressed the "old man" Harvey in his first dissertation, *Against the Thesis of Harvey*:

> Thou hast observed a sort of pulsatile heart in slugs, flies, bees and even squill-fish. We congratulate thee upon thy zeal. May God preserve thee in such perspicacious ways. But why dost thou say what Aristotle would not admit of small animals possessing a heart. Dost thou declare, then, that thou knowest what Aristotle did not?

It was as though in Harvey's book, which Primrose considered the height of useless disingenuity, learning had lost its point! Disputations, subtle argument, the charm of novelty, the desire to be conspicuous were all the well-known vices of academia. They were "frequent games" played by men of little intellect who hid within the universities, and he disdainfully rejected Harvey's new system out of hand as irrelevant and undeserving of anything more than a couple of weeks of attention.

First of all, Primrose argued in his 1630 treatise *Exercitationes, et animadversions*, which was dedicated intentionally to Harvey's patron Charles I, all veins did not have valves. That fact, by itself, could not justify the conclusion that blood in all veins flowed toward the heart. And, if blood really did circulate and if arterial blood was constantly being forced into the veins by a powerful "impetus" from the heart, then *all* the venous valves should be useless. Moreover, the heart drew only enough blood from the liver to fill its right-sided chamber. That amount was small, but its volume expanded with the heart's innate heat so that, just like milk, it experienced up to an eightfold increase in volume, as Aristotle had pointed out. Why shouldn't a small amount of blood, likewise, expand to a larger volume given the heart's high temperatures? The narrow exit offered by the heart's valves confirmed that no more than a drop of blood passed through at each beat, which, he guessed, amounted to about one ounce every two hours. "As for Aristotle," he reiterated, "he made observations on all things, and no one should dare contest his conclusions." The inability to see the septal pores at dissection he explained away as Galen had done; they adhered together after death. In any case, the pores had been accepted by the eminent botanist Gaspard Bauhin as well as by Jean Riolan, his revered mentor at Paris. Only Colombo had failed to see them and devised an unnecessary and unlikely passage of blood through the lungs!

At first blush, given Harvey's sound thesis, such arguments should seem trivial. But there was a deep and dangerous aspect that lay waiting for the unwary. The human tendency to distort provocative scientific conclusions was becoming clear to Harvey. It was equally transparent that Harvey had provoked the clergy, although, for the moment, he knew neither who would follow in the lists against him nor what they would write. Between the first chill of realization and the eventual recognition lay a long shadow.

Once the keystone was kicked aside, an avalanche was inevitable. Harvey's most vociferous opposition abroad came from his own alma mater at Padua. A reprint of almost the entire *De motu cordis* with childish arguments and acerbic criticism against it, section by section, was published in 1635 by the Venetian Aemylius Parisanus (Emilio Parisano), who had been a protégé and assistant of Fabricius before Harvey's time, and who was now an important member of the Venetian College of Physicians. The book was more often self-contradictory than not, and, in the end, Parisanus found an excellent support for his objections against novelty: things had been traditionally so because nature had wished them to be! Five years later, Harvey's friend and supporter George Ent undertook the task of publicly refuting those venomous denigrations in his own publication *Apologia pro circulatione sanguinis, qua respondetur Aemilio Parisano, medico Veneto.*

To complicate matters, Cecilio Folli, a Paduan medical student, once again came across that congenital opening in the adult septum, a patent foramen ovale, which he misleadingly used to vindicate Galen's porosities. Fortunio Liceti, professor of philosophy at Padua and Bologna, also refuted Harvey's conclusions and, in a letter to the influential Danish anatomist Thomas Bartholin, proposed an alternative fanciful system of his own: transfer of blood from the right to the left side of the heart took place through the coronary veins of the heart!

WHEN HARVEY VISITED SPAIN IN 1631 IN THE COMPANY OF YOUNG James Stewart, fourth Duke of Lennox in the Scottish peerage, he presented the court physicians with copies of his book. He may have conducted vivisections to convince them. But the power of Catholicism, with its emphasis on inherited wisdom, was overtly at work. The Spaniards knew their Galen and his timeworn principles well, as they did the medieval Arabs who had transmitted Greek learning with their own speculations tacked on. The reverberations of Galenic thinking made for a less-than-enthusiastic reception for the circulation.

Harvey realized that if additional support did not come, he would be in troubled waters.

The hour brings forth the man; in this case, two emerged from an unlikely quarter: Walaeus and Sylvius, both from Holland. Interest in the circulation had been high at the University of Leiden because of the reprinting of Harvey's original text by Jean Maire (also known as Ioannis), who had included, in the same volume of 270 pages, the objections advanced by both Parisanus and Primrose. Maire was the most prominent publisher and printer in the Golden Age of Holland and published over five hundred titles, including works by Erasmus and Descartes.

One student attending the lectures and vivisections conducted at the University Garden was Jan de Wale, latinized to Johannes Walaeus. Initially a critic of Harvey's theory, Walaeus was swayed by the demonstrations and converted after trying experiments with ligatures himself. When he measured the amount of blood distributed by the heart over a unit of time in experimental animals, he arrived at figures almost identical to Harvey's. With such evidence, only one conclusion was possible.[5] Regarding the heart, however, he entertained a natural movement for that organ as a response to the phenomenon of "irritation." The expansion of the blood that was heated in the heart constituted such an irritation, he held, and the heart tried to remove that disturbance by contracting, thus expelling the blood.[6]

Another Leiden student, Francois de le Böe, latinized to Franciscus Sylvius, held a public disputation in defense of Harvey; in the second tenet, he also supported the pulmonary circulation. But he ultimately endorsed a different theory of heart motion, the Cartesian model (which we will encounter in a later chapter). He also supported the Aristotelian model of fermentation and effervescence as the motor of all living processes, including the heartbeat. Blood effervesced in the right ventricle and boiled over into the pulmonary artery, where it was recondensed in the lungs and mixed with air particles, and was then effervesced in the left ventricle, finally overflowing into the arteries.[7] The purpose of a circulation was the complete distribution of the spirits. But Sylvius ran too fast too soon to run long. Holland was not yet ready for new ideas, and the academic orthodoxy forced him to Basel. He eventually returned as professor of medicine at Leiden, where he played a major role in disseminating Harvey's ideas.

Support came as well from a Paduan graduate, Johannes Vesling of Minden, Westphalia, in his popular textbook *Syntagma anatomicum*, a modest introduction to human anatomy that gave a clear and concise description of the circulation with full credit to Harvey. In 1638, Johannes van Beverwijck, a prominent Dutch physician and the mayor of Dordrecht who claimed to be a relative of Vesalius, acknowledged the Harveyan doctrine. Ten years later, Johannes Sicman supported Harvey's thesis at Groningen at a disputation titled *Disputatio anatomica de corde*. Jacobi de Back at Rotterdam, too, "embraced the doctrine with both arms" in his *Dissertatio de corde* to which was appended an outline of Harvey's experiments. He euphorically announced victory for the new order on all fronts.

It was on Walaeus's recommendation that his English student, Roger Drake of Somerset, elected the "circular motion of the heart and blood" as the topic for his public graduation disputation in 1639, which he ended by defending Harvey against Primrose and Parisanus. However, he advocated the traditional view that blood, heated and excited in the heart, was converted into "spirit" that forcibly expanded the heart and then overflowed into the arteries, thus causing the pulse. That was surprising because at one of Walaeus's demonstrations, the force of blood had been so convincing that, on cutting the tip of a live dog's left ventricle, it had spurted out four feet, drenching the nearby spectators! Drake was also hard-pressed to find a "purpose," or "final cause," for the circulation.

Drake's dissertation stirred the wrath of the irrepressible James Primrose. Believing Drake to be simply a mouthpiece for Walaeus (and therefore for Harvey), he leapt in and thrust, not at Drake but at his tutor. An infuriated Drake parried on his return to England by publishing his defense along with a devastating commentary against Primrose, dedicating it to both Harvey and Walaeus. Walaeus, himself, chose to vent his spleen in the letter appended to Bartholin's textbook.

THE DANISH ANATOMIST THOMAS BARTHOLIN'S ANATOMICAE WAS THE first European textbook to endorse the circulation, but only in an appendix written by Walaeus.[8] Bartholin waited until all opposition had virtually settled in Harvey's favor before he wove it into the body of his very successful 1651 edition. Throughout his life, Bartholin continued to believe in the Galenic pores as well as the production of "vital spirits" in the left ventricle. In a dissected pig, he had seen a septal defect that had been large enough to admit a pea. However, there was more to the story.

After the first edition of his textbook, Bartholin embarked upon a professional lecture tour of Italy, where he met Johann Vesling, known as Veslingius, a German professor of anatomy and botany at Padua, who was Harvey's supporter but who still maintained that the pulmonary vein carried cold air mixed with blood to the left ventricle to cool it and transported fuliginous vapors in the reverse direction. In a letter dated October 30, 1642, Bartholin wrote to Walaeus: "Veslingius has disclosed to me the secret about the Harveyan theory of the circulation, a secret that should be told to nobody, namely, that the circulation has been discovered by Father Paulus Sarpi, the Cardinal and official theologian of Venice, from whom Aquapendens has also taken his discovery of the valves in the veins." Sarpi's theories were published posthumously by his ecclesiastical successor Fulgenzio *after* the publication of *De motu cordis*.

Friar Fulgenzio Micanzio, a close friend and biographer of the well-known Venetian priest Pietro Paolo Sarpi, claimed that Sarpi had discovered the "membranes" in the veins earlier through his own independent studies on many animals. Fabricius and Sarpi were friends. When Sarpi had been severely injured in an assassination attempt instigated by the pope, it was Fabricius who had attended to him and removed a stiletto embedded in his maxilla. Galileo, a friend of both, may well have taken an incidental interest in their discussions. It is possible that they may have discussed the venous valves and the movement of blood in the veins. From 1574, Fabricius taught openly about his discovery for another thirty-five years. During that long period, neither his teaching nor his book had aroused Sarpi to make any claim to priority. Sarpi was also falsely credited with priority in the discovery of the circulation.

Bartholin wrote no more about this matter, but the worm had been introduced. Walaeus believed the story because in 1645, in an enlarged second edition of his letter appended to Bartholin's textbook, he disclosed that Harvey had learned about the circular motion from Sarpi but had made the discovery his own after he had extended the experiments. That, Walaeus revealed, was the "true story" behind the circulation.

HARVEY'S INTERACTION IN GERMANY WITH CASPAR HOFMANN OF THE University of Altdorf, not far from Nuremberg, was representative of the difficulties he encountered with Galenic supporters. Hofmann and Harvey had been friends and students under Fabricius and had kept in touch after graduation indirectly through their works. Harvey renewed contact with his old friend in May 1636 when he was in the train of the Earl of Arundel and gave a demonstration specifically for him in the university's anatomy theater, which, at that time, was the largest in the world.

Hofmann knew about the valves in the veins, although he, too, did not understand their function, and he acknowledged Colombo's pulmonary transit. In everything else, such as the ebb and flow of blood, Hofmann's views were Galenic. His criticism, in his correspondence with Harvey after the demonstration, rested upon two issues: What was the purpose or "final cause" of the circulation because only then could any science of nature be complete? As Aristotle had emphasized, true knowledge of a part or organ rested on a knowledge of its function—its *purpose* or "final cause"—as much as on its structure. Secondly, how did blood pass from the arteries to the veins without going through the septum?—the capillaries were suspected but not yet demonstrated. He also doubted that the heart worked as a pump.

Additionally, he argued, as many opponents would, that Harvey had proposed a rhythmically recurrent process whereby "concoction" ("making perfect"; purification) of blood occurred in the heart during each transit of the circulating blood. *Could* blood undergo repeated concoctions as required by Harvey's theory and still remain blood? According to Galen, the "recooking" of blood was unnecessary and therefore unlikely. A single Aristotelian concoction through fermentation was both meaningful and traditional and should surely be enough to permit both the blood's formation and its utilization.

Harvey responded with patience and understanding. It was premature, he replied, to speculate about the "purpose" of the circulation unless the *fact* of the circulation was generally accepted. The circulation should not be rejected just because it did not explain everything or it upset preconceived notions. *De motu cordis* addressed primarily the presentation of the phenomenon, the establishment of the "what" before the "why." He had demonstrated beyond doubt a circular blood flow experimentally and theoretically through quantitation, and it could no longer be ignored. He admitted that its purpose was still unclear, and he had resisted the danger of attributing a final cause. Regarding Hofmann's other point, he did not yet know precisely how the transfer of blood from artery to vein took place peripherally, but he was positive that it did occur because it was necessary.

IT WAS NOT LONG BEFORE THE CONTROVERSY MOVED INTO DANGER-ous waters. For Professors Valckenburg and Otto Heurnius, colleagues at the University of Leiden and self-righteous keepers of the mysteries, it was a matter of control. Those Kafkaesque pedagogues, feeling the necessity of a temporary alliance against a common foe, shielded themselves by cunningly piercing the septum with fine stilettos before their demonstrations to convince their students of the truth of Galen's pores![9]

Such distortions were proof enough that Harvey's enemies were ready to go to any lengths. He had seen the seething sewage behind the facade of academic life. In his conversation with Boyle, who recalled it in his notebook, Harvey told him that it was always a mistake to presume to know a person's character. Just as it was impossible to judge a man's soul by looking into his eyes, so, too, it was impossible to know how someone would behave in certain situations. Human beings were not predictable, and, in a crisis, they reacted mysteriously and unthinkably. He had learned that by watching men on the battlefields of Cromwell's war and was continually learning by watching others, and himself, in the wake of his discovery.

Harvey was quick to appreciate, too, that his new theory, which spoke an inconvenient truth so effectively and formidably, threatened to undo whole libraries of ponderous folios and patrimonies of vested intellectual interests, as well as the very art of the learned coalition that brought the sterile professors their stipends. Moreover, he had to combat crowd psychology. He realized only too well that for the generality of academics to whom the affirmation of the new and the unusual brought sudden and profound trouble, their sense of comfort and security lay within the anonymity of a faceless crowd. Honest principles counted for nothing. What ruled was the instinctive sense of belonging to the "group."

In 1637, when Harvey and his friend, the Dutchman Sir George Ent, visited Naples, they met Professor Marco Aurelio Severino. Severino was almost an exact contemporary of Harvey. Born in the small Calabrian village of Tarsia, he obtained his degree from the still-famous medical school at Salerno and completed most of his studies at nearby Naples. When the Neapolitan chair in anatomy and surgery became vacant, Severino was requested to fill it. He was the Italian who took Harvey most seriously.

As a medical academic of considerable celebrity, he was an important member of the seventeenth century's European intellectual community, and he counted among his friends Thomas Bartholin, Olaf Worm, Jean Riolan, Caspar Hofmann, and now Harvey, who, after accompanying the Earl of Arundel's diplomatic mission to Prague, had made a detour to Rome and then to Naples to meet him. Severino was a voluminous writer, and his correspondences played a vital role in the dissemination of scientific opinion. Some of his academic works went through repeated printings. As one of the leading lights of Italian medicine, he was inevitably drawn into the circulation debate.

Severino's motives for initially opposing the English doctor were empirical. But he was converted once the superiority of the discovery was demonstrated, and by 1640, he was firmly in the Harveyan camp. The Harvey–Severino friendship was a reciprocal and fruitful one.[10] Upon return to England, Harvey dispatched to Severino the only known presentation copy of the first edition of *De motu cordis*. Severino composed a lengthy critical dissertation on the subject, which he periodically modified and distributed, but never published.

In a comprehensive text on comparative anatomy, Severino emphasized the heart as the *vitae principium*. While he was not explicit about the anatomical path of blood flow, he did describe it as a circulation. In another posthumously published monograph on the respiration of fishes based upon his own studies of the gill arches, Severino's debt to Harvey was more clearly delineated. He followed the Roman historian Pliny in arguing that gills were analogous to lungs and that both absorbed air, which served as a cooling agent as well as fuel for the innate vital flame. Severino still preferred Aristotle's analogy of boiling milk to characterize the vitalization of blood in the ventricles instead of Harvey's concept of the heart as a pump.

It was not until 1636 that another favorable reference came from Italy in verses written by Johanne Rhode in honor of Harvey's colleague, George Ent, who was celebrating his Paduan graduation. Harvey was also defended "strenuously" in Rome by the physician and surgeon to Pope Urban VIII, Giovanni Trullio, who had managed Galileo's cataracts and glaucoma. In 1643, Claudio Berigardo (actually a Frenchman named Claude Guillermet, seigneur de Beauregard), an opponent of Galileo, endorsed Harvey's forceful contraction of the heartbeat and the pulse briefly in his own book, *Circulus pisanus*, but remained silent on the circulation.

To an ardent Galenist like Oluff (Ole) Worm, Bartholin's father-in-law and professor at Copenhagen, a circulation was not easy to accept. He had thought deeply upon the subject after he had received an enthusiastic letter from Leiden as early as 1631 from his Danish undergraduate student, Jacob Svabe, who had just read Harvey's book. It said:

> This doctrine so greatly impressed my mind that, for a full week, I was quite heart-sick owing to these profound thoughts. Hardly being able to calm myself by my own efforts I revealed the whole matter to an industrious student of medicine of the name of Conrigius, who is a friend of mine. Having been shown Harvey's dissertation he explained the circulation of the blood so admirably and plainly, that he himself almost seemed to be of the same heretical opinion.[11]

Conrigius (Hermann Conring) found it heretical but fascinating. When he became professor of philosophy at Helmstadt in 1632, he was among the first to defend Harvey in German-speaking countries. Eventually, Worm came around to the same question: *Cui bono?* What was the good of it all? What was the end that a circulation served? What could be the *purpose* of a circulation?

Bewilderment begets curiosity. Harvey, Worm argued, had proposed that a vital peripheral link between artery and vein, which was so essential for a circulation, took place either through direct anastomoses or indirectly through porosities in the flesh. But the former, said Worm, were still hypothetical, and the latter would cause blood to coagulate, precluding any further motion. Moreover, arterial and venous blood were so different in substance, color, and other properties that it was impossible to believe that they were the same fluid circulating perpetually.

And even if it were acceptable that the heart might perfect blood, why must the blood that had become corrupted in the tissues be perfected again? When something corrupts in nature, it withers, ages, and finally passes away. Perfecting what had been perfect over and over forever as in a circulation was so unnatural. It was the same argument that

Caspar Hofmann and many others had offered. In a series of carefully planned reasons that he published in 1632, Worm, a stickler for punctilious facts, announced that a circulation as proposed by Harvey was inconceivable.

And European professors, always eager to vent their spleen upon their rivals across the Channel, set about to make hay with such arguments against the new treatise.

HERE WAS A BITTER MISSIVE INDEED! HAD THERE NOT BEEN THINGS accepted in the sciences of which the causes had not been known, yet there was no doubt of their existence? Should we appreciate the beauty of nature less if it were less harmonious? Are not wonder and knowledge both to be cherished?

To Harvey, his thesis appeared to exude the banality of the self-evident. By what means does man judge the reason or folly of nature? How was it possible for rational thinkers to deny the value of experiment and its foundational status in the matter of fact? There is nothing so given as a matter of fact, where the solidity and permanence of fact reside in the absence of any action of human agency in their coming to be. He had revealed a reality of which the world had been in ignorance.

twenty-five

The French Connection

*But Herveius, to my way of thinking, has not been
successful in dealing with the movements of the heart,
for he imagines, against the opinion of other physicians, and against the common judgment of the eye.*

—René Descartes

J UST BEFORE DECEMBER 1628, THE FRENCH MATHEMATICIAN FA-
ther Marin Mersenne presented a copy of *De motu cordis* to his
friend, the abbey Pierre Gassendi. Mersenne and Gassendi were almost
contemporaries. The former was born in Maine, France, the latter in
Provence. The two slightly resembled each other, with their regular
features, small statures, and frail bodies. Both were vegans. Mersenne
spoke in monotonous tones and there was a certain lack of vigor in his
deportment. Gassendi, on the other hand, seemed livelier, his countenance more frequently illuminated by a slightly mocking smile. Both
had been scholars before they donned the attire of the Order of the
Minims at the same time as did the French chief tax officer Etienne

Pascal, who liked to take his teenage son Blaise to the conferences of savants at Mersenne's convent, which Thomas Hobbes also attended briefly. The order had been founded over a century earlier by Saint François de Paule, who had chosen the term *Minims* to humble the order beyond the Franciscans, who had called themselves the Minors.

Both Mersenne and Gassendi started their careers as clergymen and teachers. Their first books, too, appeared almost simultaneously and dealt with timeworn themes such as Genesis and Aristotelianism, with animated attention to ephemeral details and unique occurrences. Both evolved beyond their early interests to see the value of experiment. For both, their passion for science equaled or surpassed their passionate interest in metaphysics, and they now ranked themselves among the new mechanistic scientists who opposed Aristotelianism. Father Mersenne's death was provoked by a surgeon who, with usual masterly incompetence, severed the artery in his right arm. Gassendi, likewise, died from excessive medical bloodletting.

Both men knew *De motu cordis* well. What pleased them most was that Harvey had dared to contradict Aristotle. When Harvey met them in Mersenne's cell in the convent of the Order of Minims at the Place Royale in Paris, they spoke warmly of his book. There was one thing, however, that Gassendi could not accept—that no blood at all trickled through the interventricular septum. Surely, he persisted, there were not only invisible porosities but even open pathways were certainly present. How else could one explain the vivid and convincing demonstration he had just witnessed?

That specific event had occurred during a public dissection by Sieur Payanus, "a cunning surgeon of Aix," who had negotiated a probe through the septum and thus proved the existence of pores. At that demonstration, Payanus had taken a spatula and attempted to probe, not straight on as others had done—he first approached the septum from the right side with the edge of an iron instrument and then explored it further by twisting the instrument up and down and to the side. A penetration had at last been achieved into the left-sided chamber. The onlookers cried foul! Payanus had forced the break himself! The surgeon responded by requesting some viewers to step forward and, using a sharp knife, he cut the septum down to the instrument, and had so demonstrated a passage or canal covered with a smooth membrane (most likely a small septal defect, occasionally found in adults). The audience had gasped audibly. There *was* a definite passage within the septum! And since one was present, it could not be without use.

That was the worm in the apple, Gassendi confided quietly to Harvey. He advised Harvey that his otherwise erudite theory should be modified to acquiesce that at least some percolation occurred through such septal passages.[1] Otherwise, an unexpected gust from some passing freak wind would mess up the papers on his desk, and he would see his world vanish around him and beneath his feet, immediately and instantaneously.

Time would show that Harvey was, in fact, not sheltered from such sudden gusts.

ALTHOUGH RENÉ DESCARTES MAY HAVE HEARD OF DE MOTU CORDIS from Mersenne and may have even received a copy when he visited Paris in the winter of 1628, it was not until the fall of 1632 that he read the book.

Unlike Harvey, Descartes is an extremely sympathetic figure. Born at La Haye (which is now called "Descartes"!), a small town in Tours, he studied law at Poitiers and began life as a soldier in the Dutch military academy, later battling in Bavaria "without fire in his belly or blood on his sword." He did not like his commanding officer, and obedience was irksome as well, so he obtained his discharge. He wrote a book on fencing and even fought a duel over a beautiful lady. But he soon discovered that all he wanted to do was think. He never married—he never found a woman whose beauty was comparable with that of truth! That was no transient willfulness but a passion as vehement and fiery as first love—a deep-seated, earnest, ineradicable longing and the very purpose of his inner being.

The art of making one's life free and independent is a difficult task where a thinker is concerned. Thinking was not an independent vocation, and a thinker had to seek favor of the mighty or become the protégé of a gracious master or hunt a sinecure here or a pension there and be content to sit below the salt and count himself no better than an artist. If he did not want to starve, he had to write flattering dedications to the vain, frighten the timid with virulent pamphlets, wheedle money out of the wealthy with begging letters, and wage incessant and undignified social warfare to secure his daily bread. Generations of thinkers and artists had lived from hand to mouth until Beethoven, who would be the first of the great creators to demand his rights as an artist, and the first to exercise those rights ruthlessly. Very early, the composer had seen through the illusion of mundane society.

WHEN DESCARTES READ HARVEY'S BOOK, HE SMILED. IT INTERESTED him immensely because it was about a topic on which he, too, had been working, and it was, like Descartes's own account, a clear break with tradition. His new "Cartesian" philosophy required a physiology that was, in principle, independent of inner experience and was thus, in the modern sense, "scientific."

Even before he read Harvey, Descartes had come to similar conclusions, not from experiment but from thought, based upon classical and contemporary doctrines. He had dissected animals and had taken residence at one time on Kalverstraat, which means "cow street": a street of butchers. "There was one winter in Amsterdam," Descartes recollected, "when I went almost every day to the house of a butcher to see him kill the animals and to have carried to my lodgings the parts that I wanted to anatomize more at my leisure." Anatomy was in vogue at that time, as evidenced by Rembrandt's choice to paint *The Anatomy Lesson of Dr. Tulp*, which shows a human cadaver being dissected in an amphitheater. The dissections, commemorated by that painting, were an annual event at Amsterdam, performed in a theater before a paying audience. That "ritual" was followed by an ample feast.

Descartes had dealt extensively with the structure and function of the human body in his own work, *Traité de l'homme* (*Treatise on man*). It is regarded as the first modern book devoted to physiology that attempts to explain the vital workings of the body on mechanical grounds. Descartes personally supervised its illustrations. In his *Discours de la méthode* (*Discourse on the Method of Rightly Conducting One's Reason and of Seeking Truth*), he gave prominence to Harvey's theory, stating forthrightly that his own anatomical work had led him to the same conclusion: blood did indeed circulate around the body. He latinized Harvey's name to "Herveius," and he acknowledged the priority and precedence of the Englishman with perfect candor:

> I have no need to add anything to what has been written by an English physician to whom we must all render a homage of praise, since he was the first to break the ice in this matter and the first to teach that there are several small passages at the extremities of the arteries by which the blood which these receive from the heart enters into the little branches of the veins through which it moves at once towards the heart, so that its course is nothing else than a perpetual Circulation.[2]

The influential *Méthode* popularized Harvey's discovery in France and throughout Europe. Indeed, it would be true to say that the initial reception of Harvey occurred essentially under Cartesian influence. But the book disagreed with Harvey's explanation of heart motion. Descartes could not entertain the idea of that organ acting as an involuntary muscle, less so a pump, since he held that all muscles were controlled by the will under the influence of the mind. He refused to believe the heart as the autonomous motive force behind the movements of blood.

THAT THE HEART WAS "A VERY STRONG MUSCLE" HAD BEEN KNOWN since the days of Hippocrates. Aristotle had explained its movement in terms of the general movement of muscles, categorizing it as a "natural" movement (an involuntary muscle, as we say today), for, although it was a muscle, it was not controlled by the will. Erasistratus, at Alexandria, may have been the first to infer that the heart functioned as a double pump, a "two-stroke" pump with a double action that combined a negative force (suction) and a positive force (propulsion) to move the blood and pneuma. Perhaps he was inspired by the operation of a pump with two alternating sets of valves invented by his contemporary Ctesibius, but it is more likely that he got the idea from a smith's bellows.[3]

Harvey, too, would be impressed by another newly invented pump, nine years after the publication of *De motu cordis*. The first description of a mechanical pump that might have suggested an analogy was published in 1615 by Salomon de Caus, an engineer in the service of the Prince of Wales. It was incorporated into a fire engine to rescue burning houses. Since Caus and Harvey both frequented the court circle, it is possible that the two may have known each other, and Harvey may have learned of the pump well before his own publication, either directly from Caus or from his manuscript. When Harvey returned to London in 1637 after his European journey with the Earl of Arundel, a fire broke out near Arundel House, and one of Caus's new engines was used with such great success that Charles I authorized the Lord Mayor to place an order for many more of the same. That Harvey had heard of the fire, and perhaps may have witnessed it, is not impossible. The new engine made a stir in London, and it was most likely on that occasion, his friend Sir Francis Glisson commented, that Harvey confirmed the analogy between the heart and a mechanical pump: "This spouting out of the blood from the harte may be compared to the casting of water by the engine of late invention for the quenching of fire."[4]

More than a decade earlier, Harvey had concluded that the heart served as an impellent, or "impetus" using Aristotle's terminology, to propel blood out of the organ. Arterial filling by "impetus" was an important argument against Galen, who had reported that arteries filled by suction like a bellows when, and because, they were *actively* dilated. Harvey showed they were filled passively, like bladders that were forcibly opened by the intrusion of fluid pressed into them. In his lesser-known work, *De motu locali animalium* (*Local movements of animals*), Harvey furthered the general concept of an elementary capacity for perception and irritability by living tissues. On that basis, he explained the triggering of heart muscle contraction through the "perception" of the inflowing blood. Thus, the cardiac rhythm arose from an "irritation" between the heart and blood.[5]

HARVEY'S CIRCULATION ATTRACTED DESCARTES BECAUSE A SINGLE site, the blood in the heart, could be the source for all other motions in the body. There was no need for complex secondary causes to explain blood flow, like Galen's "attraction" or the "natural faculties." It was necessary for Descartes that the heart be the sole site and source of animate motion. It was equally vital that it should not generate its own motion.[6] He had already derived the heart's motion from a major principle of his own natural philosophy, namely, heat, which he viewed as a purely natural phenomenon.

The ancients, Descartes pointed out, had noted that living creatures were warm, regardless of environmental temperatures. The Roman senator Cicero, quoting a Stoic doctrine of the Greek Cleanthes, had asserted it as "a law of nature that all things capable of nurture and growth contain within them a supply of heat without which their nurture and growth would not be possible . . . this element of heat possesses in itself a vital force that pervades the whole world." There must be an "animal fire" within the living body that the organism replenished. That living flame was untouched by any physical agency and was an innate quality, or natural property, of the body and was connected with life.

Using an analogy introduced by Galen of the heart and a lamp, Descartes assigned heat as the prime cause of vital movements, including the heart's. It fitted well with his mechanistic theory of energy, which involved a conversion of heat energy into mechanical energy (motion) that accounted for both the warmth and the motive properties of blood. What governed the living and the dead were the absolute laws of nature. Descartes actualized those fundamental principles through a conceptual reinterpretation of "vital heat" as a physicochemical re-action, and of the "vital spirit" as a stream of particles. He held that the constant heat within the heart served as a sort of fire, nurtured and sustained by blood which was a combustible and readily expandable fluid whose particles the heat threw into a violent turmoil, a process he described as "effervescence." Through the subsequent friction and bubbling of the particles, a subtle "spirit" was released. The relation between respiration and "vital heat" was a reaction of chemical com-bustion. Such a "fermentation theory" of animal heat had been pro-posed nearly a century earlier by Vesalius's teacher, Jacobus Sylvius, and would be again by the Flemish chemist Van Helmont, who was Harvey's acquaintance. Sylvius's heat arose from the effervescence of humors when the products of digestion (the chyle) mixed with blood. Van Helmont speculated that heat stemmed from the blending of sulfur and volatile salts present in blood.

To Descartes, the heart was a kind of furnace that supplied energy by virtue of its natural heat, which, in turn, caused blood to expand. The essential motion of the heart, therefore, was not the contraction of its walls, as Harvey had ascribed, but an Aristotelian *expansion* that resulted from this heat, which was generated within the left heart cham-ber. The visible change in the blood's color within the heart was evi-dence of that vital flame. Heat was so vital to heart function that, in a letter to Mersenne dated February 9, 1639, Descartes wrote, "I am quite ready to accept that if anyone should think that what I have written about this [the mechanism of the heart's movement] should prove false, all the rest of my philosophy is of nothing worth."[7]

DESCARTES DID NOT ALWAYS THINK AS CORRECTLY AS HE BELIEVED HE did. In this instance, he envisioned the expulsion of blood as a drop-by-drop process rather than a substantial flow, as Harvey proposed. The physical laws of his corpuscular natural philosophy and his mechanical view of animals as living organisms that conformed to the same laws as external nature *required* motion to be transmitted particle to particle.

Harvey's doctrine, which described a forceful spontaneous contraction, was anathema.

Cartesian physics envisioned the movement of particles to always be in the form of vortices. The cosmos was filled with fluid vortices of particles that caused the planets to circle around a central star (the sun). So, too, it must be with the microcosm. The circulation of blood was no autonomous whole movement generated by a muscular heart, as Harvey had envisioned, but rather a chain of forced movements, a "vortex," of the particles.[8] From the left auricle, Descartes explained, the left ventricle received blood drop by drop that, becoming mixed with the ferment within that chamber, induced an effervescence, or "ebullition," which not only "subtilized" and rarefied the inflowing blood intermittently but also, by its bubbling, caused the rarefied and expanded blood to dilate the heart and then flow over with "an effort and by many jolts" into the arteries, just as Aristotle had described in his analogy of hot milk boiling over. Blood became sufficiently agitated in the heart that it could force its way into the aorta and its arteries.[9]

Every heartbeat and arterial pulse were caused solely by a heat-induced blood expansion that, set into motion in that way, delivered the natural heat everywhere. The heart valves prevented blood from backing up "in agreement with the laws of mechanics." Some heated blood remained in the pores between the heart fibers and that, now mixed with blood freshly dripping in from the left auricle, served as ferment for the next heart expansion, thus resulting in a ceaseless chain reaction of "active" expansion and passive contraction as in a combustion engine. There was "no other heat in the heart but this movement of the blood particles."[10] The purpose of respiration was to cool the blood. Blood that returned to the left ventricle, drop by drop, was condensed to a form suitable to the production of ebullition, thus completing a circulation.

But "ebullition" of blood brought quantitative problems. It was difficult to measure the amount of blood that fermented as it passed through the heart—a fact that Primrose and Caspar Hofmann had recognized. Fermenting blood left the heart in small and irregular amounts and was therefore measurable only in extremely small units, if it could be measured at all. Ultimately, Descartes was satisfied with his mechanistic explanation, and though he endorsed completely the concept of a circulation, he remained opposed to Harvey's radical doctrine of the heart as a pump.

Harvey refused to accept Descartes's explanation that heart motion was due to a fermentation process. Above all, he protested, fermentation was not strong enough to produce the noticeably vigorous and regular beats of the heart. Only subtle, irregular motions could be induced by fermentation, and they would differ markedly from any observed heartbeat. He was also not impressed with Descartes's theoretical classification of the categories of fermentation in which one type had been designated as being quick enough to produce the heartbeat.[11]

DESCARTES SUPPORTED HIS VIEWS BECAUSE HE HAD FELT THE LIVING hearts of animals with his fingers and found them to be warmer than the rest of the organism. Moreover, an excessive amount of heat was not necessary for rarefaction or even for fermentation. In further support, he presented two experiments, which, paradoxically, had the singular effect of refuting his own theory!

In the first experiment, a heart was taken out and deprived of all its blood and shown to continue beating. Harvey was quick to point out that such pulsations could not be ascribed to a heat-induced expansion of blood since there *was* no blood in the extracted heart chambers. Descartes objected. He introduced another analogy derived from log fires. The empty heart muscle with its natural heat, he claimed, was still saturated with a few particles of blood within its tissues. The situation was akin to the burning of a green piece of wood that was not overtly wet but that was still impregnated with drops of moisture. When such a log burned, vapors emerged through narrow crevices in the bark and then escaped. Their point of escape could be clearly marked since the bark swelled and showed a series of fracture lines. Once the bark had broken at a certain point, the swelling disappeared because the trapped moisture had been released. Then, as new vapor was formed, the bark arched again, intermittently releasing more vapor through the same opening. This phenomenon of expansion and release induced by the vapors occurred rhythmically and was comparable to the pulsation of an isolated heart.

In a second experiment, Descartes introduced a finger into the heart chamber through a hole at its apex and was able to feel the organ's strong contraction. Again, the conclusion was obvious to Harvey, but Descartes gave a different twist. The finger, he explained, was not squeezed by the walls of the ventricle but by the tendinous fibers (the chordae) attached to the papillary muscles of the heart that were being distended during ventricular *dilatation*.

DESCARTES'S FIRST EXPERIMENT WAS DUPLICATED BY A CONSERVAtive Dutch physician Wilhelm Plemp, a professor at the University of Louvain. Plemp demonstrated that on removing the living heart and dividing it into pieces, even the fragments continued to beat. Thus, the effervescence of blood in its chamber, as Descartes had contended, could *not* possibly be the cause of the heart's motion. Moreover, such fragments of heart tissue were not even hot enough to vaporize blood.

Descartes responded promptly. Every heart, he explained, was predisposed from its embryonic origin to pulsate, and in the adult did so with so much facility that the very minimal possible fermentation ensured its continuation. In the heart pieces, a mere drop sufficed to continue its pulsation for long periods. Such tiny drops were activated by a ferment that was hidden in the recesses of the heart's substance in every cut-up fragment. This "fermentation theory" explanation was quite different from Descartes's earlier account of a forceful diastole of the intact beating heart. But it was necessary for Descartes to formulate a satisfactory response to Plemp's criticism since Plemp was well placed in a good academic position to disseminate the Cartesian natural philosophy within the university. If only he could be convinced!

Descartes's innumerable experiments, with their wide range of results, offered him many options to pick and choose, at will, those that suited his immediate needs. In one of those, he told Plemp, he had excised the heart of an eel and waited until the heart had clearly died. Then he had brought it back to life, and contracting again, by just warming it and injecting into it some drops of blood that he had kept aside for the purpose. Plemp was unimpressed. Whereas, at one time, Plemp had fitted himself snugly into that middle world in which men take no sides in conflicts and decide no great causes and make no great refusals, by the second edition of his book *De fundamenta medicinae* he had been won over by Harvey, not by Descartes. It was precisely that part of Harvey's doctrine that Descartes had rejected, for his own reasons, that Plemp found the most convincing. Plemp became an advocate of the circulation in the way Harvey intended, and turned out to be "more royalist than the King," so to speak, more pro-Harvey than Harvey himself!

Undoubtedly, Descartes, in his youth, had posted a brilliant academic record. But he had also displayed a sharply combative, if not belligerent, temperament. It was clear that Descartes's experiments were designed only to confirm his own logically preconceived requirements. When certain investigations seemed to contradict his ideas, Descartes had no qualms to doubt such results, because he maintained, as Plato had, that observations through the senses could, by themselves, be confusing and ambiguous. Only those conclusions that could be *reasoned* clearly were the truth. Moreover, he was quite at ease with falsifying for his own advancement! In contrast, Harvey based the certainty of his knowledge primarily on experiment, which he placed side by side with reason. In so doing, Harvey remained completely within the Aristotelian framework of thought, even though he may have contradicted individual tenets of the system when he identified new phenomena through more precise observations and better experimental techniques.

DESCARTES, WHO WAS A FREELANCER OUTSIDE THE UNIVERSITIES, needed an agent within to yoke his new medicine to academia and widen the membership to his new philosophy, including focusing the minds of novice students on matters Cartesian. And, disappointed with Plemp, he found one, at least for a while, in Hendrik de Roy, known as Henricus Regius, a professor of botany and theoretical medicine at Utrecht, who became Descartes's first true disciple. Besides, Regius came from a rich family of brewers and so had the town fathers behind him as well. Descartes suspected that Walaeus was being used by a wily Harvey in the same way that he, himself, was manipulating Regius! In fact, to many, Walaeus was now the principal protagonist of the circulation.

Descartes was delighted with Regius. Here was a grateful pupil with a voice in a university. On his side, Regius believed that he owed his professional advancement to Descartes. Under the master's tutelage, he began to institute provocative disputations on the circulation in his own academic curriculum. He was not the only one to do so, as Walaeus at Leiden was also disputing in favor of Harvey's doctrine. When Regius sent drafts of his lectures to Descartes, who was equal to Odysseus in cunning, the philosopher edited and carefully monitored them so that Regius presented only what was in Descartes's own interest.

Descartes was precise in defining the processes of rarefaction and effervescence in the heart to keep the action and function of the heart truly Cartesian. He was also insistent on not mentioning Harvey or Walaeus in any lecture. The final discourse of the series was a triumph because Regius put on view how the idea of a circulation, albeit drop by drop, was central to Cartesian physiology and physics. But, almost from the beginning, Descartes had found Regius difficult to control. Gratitude does not necessitate admiration. When the disciple went astray and began to advance his own contrarian views, an outraged Descartes, more splenetic than becomes a philosopher, reacted with sharp condemnation and broke off with him, kicking him out of the coop.

From 1640 onward, Regius took Harvey's side, especially in the disputes against Primrose. But Regius's theory of cardiac motion would remain Cartesian. Not only was the motion of the heart, he believed, triggered by the fermentation process that he called the "fire without light," but also he associated the cause of cardiac activity with mechanically conceived circulating animal spirits that streamed into the heart fibers—a reinterpretation of the ancient innate heat that had proved to be so rich in consequences for the Greeks. He concluded, "There is no reason to attribute to the heart any force, be it one of attraction, or magnetism, or a contribution to nutrition, or anything else unintelligible or superfluous."

IN 1637, DESCARTES MET CORNELIUS VAN HOGHELANDE, A NEIGHBOR and practicing physician at Leiden. He had borrowed medical books from Cornelius to expand his ideas on natural philosophy into medicine, and the two developed a lifelong friendship. When Descartes left Amsterdam, he entrusted his trunk of letters and papers to Hoghelande. The great admiration that Hoghelande had for Descartes, "the shining light of our land" as he had once greeted him, was apparent from his 1646 tract on mechanical physiology, which reflected Descartes's doctrine closely. Holland would become a citadel of Cartesianism.

Heart motion, Hoghelande wrote in support of Descartes, occurred through fermentation by spirits, a chemical reaction analogous to that of spirits of niter or oil of antimony. However, he also praised Harvey. He offered a compromise for the "pumping" function of the heart by simply letting fermentation occur a second time on the surface of the organ within the coronary arteries! The heat of that second fermentation made the heart muscle contract. The heart's action in situ thus had two different causes: systolic (from external fermentation outside the heart cavity) and diastolic (an internal Cartesian fermentation within the heart chamber), whereby Hoghelande continued to consider diastole as the primary action by which blood left the heart. The work did not remain completely unread. Thomas Bartholin's influential textbook listed Hoghelande as well as Descartes as those who had established the "ebullition of blood in the heart."

The disagreement between Harvey and Descartes concerning the heart's motion did not arise because of observational discrepancies but because of their differing interpretations of the *same* phenomena. They presented two alternative models of the human body: a Harveyan mechanical one based on hydraulics and a corpuscular Cartesian one based on chemical processes. The real issue was not one that could be seen or perceived, nor on facts as might be demonstrated by experiment, but was how those events should be decoded. The misunderstandings and contradictions were like the disjunction between the recto and the verso of a page, which never meet even though they are the self-same thing. Given the heat of the debate (no pun intended) and the stature of the opponents, it was an intense moment in the history of science. Physiology was at the intersection of what could be regarded as a horizontal–vertical crossroads: the point at which the motion of the heart as the result of fermentation was poised against the ideal of a powerful and all-important muscular organ.

twenty-six

Bad Blood

Do not complain of the mean and petty, for regardless of what you have been told, the mean and petty are everywhere in control.

—Johann Wolfgang von Goethe

A T A DISPUTATION HELD AT THE SORBONNE IN DECEMBER 1642, it was concluded that blood did indeed move in a circle. Three years later, the question was raised whether Galen's account of blood flow should be formally revised. The president at that event was Jean Riolan fils, son of Jean Riolan of Amiens who, for a period, had been dean of the Faculty of Medicine.

The younger Riolan had inherited little from his father except his hypochondria, self-pity, and snobbish conviction that the Riolans were descended from a noble and ancient family. He was a reserved and serious lad at school, somewhat melancholic, a studious but lying little hypocrite who enjoyed pointing out the faults of others and seeing them punished. He studied intently and learned quickly, but the little

221

boy had been burdened with the fear of God's awful judgment and prayed devotedly at bedtime that his name be written forever in the Book of Life.

After graduation, he plunged into the study of medicine and demonstrated a veritable fever for acquiring and hoarding knowledge. He studied the botany of herbs and the science of unguents. He became an expert in the cure of agues as well as contusions, wounds, and abscesses. He elaborated on giants and hermaphrodites. He even studied languages. It seemed to that young man that life had but one purpose, which was to study. He reflected "not the man who works that he may live, but as one who is bent on living only in his work." Clearly, his striking future inability to produce a logical argument through his unstable capricious brain could not be ascribed to deficiencies in his education. But the boy was to blame; he was constitutionally perverse and sullen. He had neither patience nor tact enough to overcome the difficulties of a natural antipathy. Still, he had been warned: let it always be remembered that life is short and knowledge endless, and that many doubts are destined not to be cleared.

THE LIVES OF HARVEY AND RIOLAN OVERLAPPED. THEY WERE BORN A few months apart and died in the same year. Harvey remembered Riolan well—a creature in a sober black habit, more corpulent than his portrait indicated, with a scornful mouth shut tight as befits a man with nothing more to learn. No warmth radiated from his countenance. Riolan was a blighted, supremely unhappy man. He had a horror of wit and laughter as they distorted the man made in God's image. The look in his "little Eies" was not forgotten. They were black—full of light, full of darkness—all intimidation as well as sweetness and were encountered unwillingly and for no longer than necessary. That gaze was dangerously observant and, in the way he blinked, he seemed to form judgments of men, and he clearly disliked them all. Harvey recalled how those eyes had caused him to shudder: eyes like those of a priest listening to a confession.

Riolan could never like things as they were. But luckily for things, he could equally never decide how they should be. Cold and anemic, trusting neither himself nor others, he exercised his will mainly in trifling matters, painfully conscientious and attentive to petty detail. Like a poet of details, he loved making small decisions and embellishing trifling points—"that microscope of Wit / Sees hairs and pores, examines bit by bit." He was incapable of either analysis or synthesis since the connecting links invariably escaped him. He was the god of small things, maybe out of fear, or from some other self-righteous reason, perhaps to compensate himself for his uncertainty in greater matters. He regarded himself to be an oracle whose most trivial remarks carried the authority of weighty decisions.

Riolan vented his most powerful polemics against Harvey's circulation in Book III of his new publication, *Encheiridium anatomicum et pathologicum*, which came out in 1648. It was a malicious attempt, enveloped in compliments, to destroy Harvey's entire work, which, he claimed, threatened to overturn the foundations of Galenic medicine. Riolan desired Paris to be the new Cos and he a progressive Hippocrates, and he shared Sylvius's nationalistic and religious sensitivities. It was also apparent that the French professor had not, like many others, taken the trouble to read Harvey's work closely but had instead responded through gut-level revulsion to what he assumed Harvey would be saying, just as scholars earlier had done and would do since. Disappointment and distrust filled his quill. Pouncing upon the smallest details, and no detail was too small for him, this latter-day Hippocrates of the modern-day Cos asserted that Harvey's thesis could not, in any form, be accommodated within Galen's physiology. It was necessary that every doctor at Paris should defend Galen.

RIOLAN DID NOT COMPLETELY DISAGREE THAT SOME CIRCULAR trickle of blood did occur. It was the large waves of circulating blood that Harvey had described that his remorseless pedantic conscience could not accept.[1] He endorsed the Cartesian view of his kinsman that the heart, with every motion, ejected only one or two drops of blood at the most. During some contractions, no blood at all was ejected. A single drop of blood, Riolan insisted, could swell, foam, and evaporate sufficiently to fill the blood vessels so that blood took a full two or three days to circulate completely through the body, depending upon the amount of food ingested and excrement voided. Even if one considered that the heart propelled just a drop or two of blood with every contraction, there would be so many drops that it would be difficult to accommodate them all in the blood vessels! Anyway, what could be the use of a rapid, excessive flow as described by Harvey? The more rapidly blood circulated, the less time it had to adequately extract "that sweet liquor, that vivifying nectar" (what lyricism!) to nourish and warm the tissues. Riolan used just the sort of quantitative reasoning to tear down the circulation that had led Harvey to build it in the first place.

Riolan insisted there *were* Galenic pores in the septum that enabled the ventricles to communicate. Nor did he hesitate to make blood ebb and flow, or even stream backward. With an intellect of the narrowest type, dull of imagination and impermeable to ideas, he took the ridiculous and impossible position that while an antegrade Harveyan trickle circulated in the large vessels (the circle of his scheme involved only the major portions of the aorta and vena cava joined by anastomoses), blood moved in the *opposite* direction in some of the other vessels:

> In order that the veins in the arms and thighs should not become drained, it is possible that the venous blood descends as I have demonstrated, in refutation of Harvey.

Thus, there were a variety of "circulations" in the body, a view that set Harvey's teeth on edge. Nor did he believe that blood passed through the lungs:

> Harvey is very learned, but when he says that the blood passes through the lung, he is going against Nature.

THE EVER-FAITHFUL PAUL SLEGEL, WHO HAD TRIED TO INTERVENE on Harvey's behalf with Caspar Hofmann, now turned his efforts to persuade Riolan, and they proved equally futile. Riolan lacked the wit to understand and the taste to value a fine conception. In March 1651, an exasperated Harvey wrote to Slegel and described what was the last of his experiments in which he demonstrated incontrovertibly, in the presence of his colleagues, that there were no porosities in the interventricular septum.

Harvey tied the pulmonary artery, pulmonary veins, and the aorta in a strangled human corpse (a crime victim who was hanged) and made an incision in the left ventricle to expose its cavity. He then inserted a catheter through the vena cava, which was tightly ligated so that no water could leak back, into the right ventricle through which he forcibly injected "the greater part of a pound" of hot water. What happened? The right auricle and ventricle became enormously distended, but not a single drop escaped into the left ventricle as observed through an opening in the latter. There were no pores in the septum.

He then inserted the same tube into the pulmonary artery, which was also tightly ligated so that no water could leak back. When water was forced in, a torrent of water mixed with blood immediately appeared in the left ventricle. The amount washed into the left chamber was virtually equal in quantity to the amount injected. The experiment highlighted once again Harvey's skillful application of ligatures of varying tightness to investigate circulatory flows. "By this one experiment," Harvey declared triumphantly to Slegel, "I have easily butchered all the arguments of Riolan."

HARVEY WAS AVERSE TO CONTROVERSY. IT WAS NOT SURPRISING THAT the acceptance of his theory, given its novelty, should be fraught with misunderstandings and intellectual perversities. But he had complete confidence in his model, which was derived from experiments and vivisections, in contrast to Riolan, who had never soiled his own hands in the examination of the facts of nature (Riolan had a queasy dread of vivisections). In contrast, Harvey emphasized in his second discourse that he had established his thesis "by appeals to sense and experiment after the manner of anatomists." Unlike Galileo, who had spoken of the book of nature as "written in mathematical language . . . triangles, circles and other geometrical figures," for Harvey, that book lay "so open and so easy of consultation . . . through the senses."

At the same time, he was exquisitely sensitive to the professional damage that senseless diatribe could inflict when unintelligent but powerful men were made to feel insecure. Rhetoric, to him, was the febrile activity of men incapable of success, who deceived themselves to hold at bay the annihilating awareness of their lack of worth and turned a blind eye to their nullity. Some people will stop at nothing to strike a blow when they are jealous. They prosper far more in lies than in sincerity. Their contagion is communicated through passion, not through reason. For Harvey's opponents, it was not hate ad hominem. It was rather a horror, a dread that the whole rotten edifice of their own convictions might collapse.

Some things brook no delay, and this was one of them. It was not in Harvey's nature to hide. Humility was not a state of mind conducive to the advancement of learning, and his combined energy and resourcefulness was just the brew for this situation. He had hitherto remained aloof from lesser controversies, not responded to anyone since the first attack by Primrose in 1630. It was 1649 now. He was seventy-one years old, and the time had come to rise in the tidal wave of opinion and sweep his new ideas into universal acceptance as "dogs will bark and belch up their surfeits . . . but we must take a special care that they do not bite, nor infect us with their cruel madnesse, lest they should with their dog's teeth gnaw the very bones and principles of truth." Those were just the type of vermin that could destroy great works.

One thing he did not lack was a proper estimation of his own powers. Few would have called Harvey a man of business, but he was very emphatically a man of action in arguments. He aimed straight for the jugular:

It is now many years, most learned Riolanus, since, with the aid of the press, I published a portion of my work. But scarce a day, scare an hour, has passed since the birthday of the *Circulation of the Blood*, that I have not heard something for good or for evil said of my discovery. Some abuse it as a feeble infant, and yet unworthy to have seen the light: others, again, think the bantling deserves to be cherished for; these oppose it with much ado, those patronize it with abundant commendation. . . . There are some too who say that I have shown a vainglorious love of vivisections, and who scoff at and deride the introduction of frogs, serpents, flies, and others of the lower animals upon the scene, as a piece of puerile levity, not even refraining from opprobrious epithets.[2]

It was vital for Harvey to present things to his audience within a larger context, the big picture as it were, because many human minds, like Riolan's, are small, and sink under the weight of heavier encumbrances.

HIS FIRST OPEN LETTER TO RIOLAN WAS A DIRECT REFUTATION OF the *Encheiridium* that Harvey had received from Riolan himself. His exacting mind was quick to home in on Riolan's weaknesses. He approached his critic with perfect courtesy and patient argument, referring to him as "the learned and skillful anatomist, Riolan" while systematically exposing his lack of learning and common sense — indeed, it would be said that the letter should be adopted as a pattern by all who feel constrained to engage in controversy with an unreasonable opponent.

A circular motion of blood did not destroy the ancient Physick, Harvey pointed out, but rather enhanced it. Specifically, Harvey showed that "the learned gentleman, in the very booklet in which he states this, appears everywhere to declare strongly the opposite." Harvey's response was a specific rebuttal, point by point. For instance, "in Book 3, Chapter 8," Harvey wrote, "How indeed could he who has repeatedly asserted a circulation through the general system and the greater, deny a circulation in the branches continuous with these vessels, or in the several parts of the second and third regions?"[3] Harvey concluded from Riolan's befuddled thinking that:

> most learned Riolan seems to me, when he says, that in some parts there is no Circulation, to speak rather officiously, than truth; that he might please most men, and oppose nobody, and that he rather wrote humanely, than gravely, in the behalf of truth.[4]

In the end, Harvey charged that if Riolan believed in a partial circulation limited to the great vessels, he must, in fact, believe in the total.

A SECOND LETTER TO RIOLAN FOLLOWED, BUT THIS TIME THE greater part of it was a supplement to *De motu cordis* and supported the circulation through further experiments. It made new claims that rested on different arguments. The academic medical world—indeed, the world at large—urgently needed a Galileo-style reality check whereby evidence was the most pressing imperative in science. Although he addressed the letter directly to Riolan, it was intended to cut the wind from many hostile sails. Adding Riolan's name allowed him to conveniently publish the two letters between the covers of a single new volume titled *The Circulation of the Blood*, which was printed simultaneously at Cambridge and Rotterdam.

Harvey had painstakingly continued his experiments for decades after the publication of *De motu cordis*. The second letter began with further proof that the arterial pulse was produced by the force of blood driven into the arteries by the heart. The common practice of bloodletting made evident that blood flow from the cut end of a vein could be impeded by finger pressure *below* the cut. The direction of flow *from* the arteries *to* the veins became immediately obvious when both a vein and artery were cut. Blood squirted forcefully as a spout from the severed artery but was nonpulsatile from the severed vein. Furthermore, arterial blood always squirted with high velocity in an abundant flow, not drop by drop as Riolan and Descartes had imagined.

To demonstrate that the swelling of the vein below a ligature was not caused by heat or ebullition, Harvey plunged the ligatured limb into cold water until it was chilled. The veins still showed distention and, since cold water would have precluded accumulation of heat, the swelling could only have resulted from a collection of blood flowing toward the heart that could not escape from the veins. Finally, when equal quantities of arterial and venous blood were withdrawn into separate vessels and allowed to clot, the two clots looked precisely the same, dispelling the idea that they were two sorts of blood.

In 1650, in one of his usual gossipy letters from Holland, Thomas Hobbes informed Harvey of Descartes's death. Hobbes had not liked Descartes. He had fought the man and his ideas, but in his letter, he avowed distress by the death.

Harvey, too, was sad at the demise of his adversary. Some adversaries can serve as a goad and Descartes, after all, *had* been among the first to declare that the circulation was correct and important. His bone of contention was the mechanics of cardiac contraction, about which he understood nothing. But he had been a good man who had defended the circulation with intelligence, notwithstanding his new philosophy which had seemed to Harvey to be more a play with words that ends by giving an illusion of the real, provided one played with them long enough. How else was it possible to doubt one's own existence, as Descartes claimed in his philosophy, except by a play upon words? Such men often went astray for *their own* truth in their search for Truth. But Descartes *had* refused to be influenced by others and had endeavored to be wholly free in his support of Harvey. As for Riolan, the two horses, judgment and emotion, ill went together, at best, for him and, at bad times, the one lay down immobile while the other was galloping.

twenty-seven

A Confederacy of *Circulateurs*

> *No one can produce a theory so sound but that facts,*
> *time, or use may not bring forth something new to show*
> *that one's fancied knowledge to be ignorance, and that*
> *one's first judgement is repudiated by experience.*
>
> —PUBLIUS TERENTIUS AFER (TERENCE)

I N 1652, A SLIM BOOK WRITTEN IN THE FRENCH VERNACULAR BY A
surgeon named Jean Martet was published at Montpellier.[1] It supported Harvey's circulation though did not mention Harvey by name; it drew its information from Walaeus, who had named Harvey with distinction.[2] The Montpellier doctors had remained Galenists at least up to 1650, when Lazare Rivière (also known as Riverus) was called upon to resign for teaching the circulation. He transferred to Paris.

When Thomas Bartholin had attended that school in 1641, he had found the Montpellier faculty still wrapped in Galen's authority. By 1648, Jean Pecquet was already demonstrating experiments to support the circulation—he is best known today for his work on the lymphatic system. Martet first summarized Pequet's work and then gave a clear account of the new circulation theory, including the experiments with ligatures, the passage of blood through the lungs, and the objections raised by its opponents.

The following year, Riverus, now adviser and physician to Louis XIII, became the first to teach the circulation formally at Paris. In 1660, another Frenchman, Normandy-born Pierre Vattier, physician to Gaston, Duke of Orleans, advanced Harvey's thesis in his published booklet of fifty-six pages titled *Le coeur dethroné* (*The heart dethroned*), in which he credited Harvey's discovery to *himself*, stating that it was "that same source of blood stream that I have been proclaiming in my hospital lectures, and elsewhere, for the past thirty years!" Pierre Dionis, anatomist and surgeon to Louis XIV, persuaded his royal master to order the Sorbonne faculty to endorse the new teaching. In his 1698 volume *The Anatomy of Man According to the Circulation of the Blood*, he attacked Descartes, and disposed of the latter's entirely false thesis regarding the mechanism of contraction of the heart. A splendidly illustrated anatomy was published in 1749 by Jean-Baptiste de Sénac to support Harvey's work: it was the influential volume *Traité de la structure du coeur* (*Treatise on the structure of the heart*), which completely routed the upholders of Galenical misconceptions. Sénac was Physician in Ordinary to Louis XV and a member of the Academy of Science.

By the end of the century, France still had a group of physicians and natural philosophers who remained opposed to Harvey. They called the supporters of the circulation *circulateurs*, a term synonymous with "charlatan," and themselves the *anti-circulateurs*. In Paris, Beretrand published a book against Harvey in 1652, and the university continued to accept graduation disputations against the circulation for at least another two decades. Gui Patin, Riolan's successor, even applied to the king to have the circulation theory officially banned in the whole country, reminiscent of Sylvius's attempts against Vesalius. Gui had just the qualities with which a viper is credited: cunning, with callousness and poison. In his petition, he claimed that the circulation was "paradoxical, useless, erroneous, impossible, absurd, and harmful!" As dean of the Faculty of Medicine and professor at the Collège de France, he had acquired his reputation more by his deliberate and aggressive opposition to any sort of novelty than by serious and considered original observation. Busy by temperament with all sorts of matters in all sorts of places, he and his son were dealers in clandestine books and were parodied as such in the plays of Molière.

THE CONTROVERSY REGARDING THE CIRCULATION WAS NOT LIMITED to physiologists. The literary men in France, the ones who held the eager ears at court and in public, propagated "pro-circulation" ideas. Primary among them was the Auteuil Literary School, so called because the group of Molière, Boileau, La Fontaine, and Racine met in what was then a village in the environs of Paris. It is remarkable that three of the greatest French writers of the Grand Siècle should, upon their own choosing, address a scientific question that one would have thought should have left them indifferent.

Molière, always on the lookout for topical subjects, reflected the popular reaction to the crumbling edifice of Galenic theory, whose cracks even a layperson could now see. The sarcasm in his play *Le malade imaginaire* was apparent:

> What I like in him is that he is firmly attached to the opinions of
> the Ancients and never will consent to understand, or even listen,
> to the reasoning and experiments connected with supposed discov-
> eries about the circulation of the blood and other matters of the
> like sort.

Boileau, Molière's friend, was a mordant critic of the *anti-circulateurs*. Finally, in just twelve lines of verse in *Le quinquina*, La Fontaine, after learning from Dionis exactly what Harvey's doctrine was, admirably summed up the discovery: "Blood, the source of life . . . circulate[s] / Through the veins, without ceasing, to the arteries . . . / This goes back again and is purified." There was another negative reference to Cartesian heart movement in his *Discours*, addressed to Madame de la Sablière, at the beginning of the first fable in Book X, "The Two Rats, the Fox, and the Egg."[3]

twenty-eight

Once More into the Breach

*Nothing that is vast enters into the life of mortals
without a curse.*

—SOPHOCLES

WHO GETS CREDIT FOR A DISCOVERY? INVARIABLY, WHEN ONE IS made, there will be passages found in earlier writers that seem to imply its anticipation. There had been priority battles before Harvey. If the expectation of reward was a good recipe, so was the prospect of success. At the opening of the modern era, Galileo, whom Milton called the "Tuscan artist with his optic glass," had faced a host of rivals. He was the only contemporary mentioned in *Paradise Lost*; indeed, it was said that, without Galileo's discoveries, Milton's world would have taken a less grandiose form. More controversies in science followed: Newton and Leibniz, Hooke and Huygens.

Harvey's own forerunners had been many, but though they had hinted at the truth in scattered commentaries, only he had proceeded forward, step by step, to clearly see the details others had overlooked.

The English mathematician Walter Warner, Harvey's contemporary, had claimed to be the "only begetter" of the circulation and even maintained that he had communicated his thoughts to Harvey prior to the publication of *De motu cordis*! There was also the warning issued by Harvey's close friend and supporter Sir George Ent in his presidential address, given in 1665 to the College of Physicians, of the danger of reading meanings into older texts that had not been intended by their authors. Indeed, by doing so, it would be possible to show that even Galen may have known that blood circulated through the body!

IN 1655, GIOVANNI NARDI ANNOUNCED THAT ANDREAS CAESALPINUS had anticipated the circulation in the summer of 1566, half a century before Harvey.[1] Caesalpinus was, indeed, the first to coin the term *circulation*, as well as name the still-hypothetical peripheral anastomoses between arteries and veins "capillaries." However, Harveyan scholar Gweneth Whitteridge has clarified that Caesalpinus applied the term *circulatio* to the alchemical process of cooling the hot blood from the heart and not to the movement of blood in a Harveyan sense. Moreover, his description was not based upon direct observation at vivisection or dissection but on largely sagacious arguments within the narrow compass of Peripatetic naturalism. Indeed, it is remarkable how close to the truth one could get in his time just through speculation! Notwithstanding, Harveyan scholar Walter Pagel acknowledges that Caesalpinus, through his many statements, did mark an early breakaway from Galen and paved the stepping-stones for Harvey.

Caesalpinus may well have vivisected living animals and exposed the pulsating arteries and the accompanying veins. The heart, he wrote, was the source of all blood in the body, both arterial and venous. He observed the consistent swelling of a vein on its peripheral side when a ligature was applied, indicating that venous blood normally flowed from the periphery to the center, and he ascertained as well that a ligated artery bulged on the cardiac side, implying a flow toward the periphery. Those ligature observations should have suggested a circular course of blood flow, but no such inference was drawn. He was too steeped in Aristotelian doctrines, even at the price of sacrificing his own correct observation of venous blood flow. Harvey described the same experiments almost word for word, but within the context of a scientific theory. Caesalpinus had also grasped the important truth that the heart, during contraction, discharged its contents into the aorta and the pulmonary artery, and during its relaxation received blood from the venae cavae and pulmonary veins. And he did elaborate upon the absurdities that could result from Galen's description of a simultaneous contraction and dilation of the heart and arteries as well as that of the heart and the lungs. But he failed to assemble his observations into a cogent hypothesis.

In another passage, Caesalpinus interpreted the flow of blood as an outward movement through the arteries and a return movement through the veins, but one that occurred only at certain times of the day, such as during the "waking state":

> If we take account that in the waking state, there is a movement of natural heat towards the exterior, that is to say towards the organ of sense, while in the sleeping state there is, on the contrary, a movement towards the interior, that is towards the heart, we must judge that in the waking state much of the spirit and blood become engaged in the arteries, while, on the other hand, in sleep the animal heat comes back through the veins to the heart, but not by the arteries since the access provided by nature to the heart is through the vena cava and not through the aorta.[2]

Within the same framework, he also compared the movement of blood to the flux and reflux of the Euboean Sea, reminiscent of Alcmaeon, the Greek natural philosopher who had begun the circulation story on the basis of Homer's wine-dark sea:

> Now when we are awake the movement of the native heat takes place in a direction outward, namely, to the sensory regions of the

brain. When we are asleep however it takes place in the contrary direction towards the heart.[3]

In another book, *On Plants*, Caesalpinus referred to a contrarian movement of venous blood *away* from the heart toward the organs, and he mentioned (though only once) a partial flow of blood from the right to the left ventricle through the septal pores:

> The blood flows from the right ventricle of the heart, partly by the median wall [septum] and partly, for reasons of cooling, through the lungs.

Such juxtapositions of statements appear contradictory today, but they may not have impressed his original readers as such. It is possible that he may have understood that blood flowed in the veins from the periphery to the center as a basic mechanism and not as an exceptional instance. But Caesalpinus did not describe an exclusive centripetal movement of blood in all veins at all times. He failed to answer how blood made it to the peripheral veins and how, in a one-way flow, blood manufactured in the liver, an idea he endorsed, reached the arteries. Throughout his works, he discussed and toyed with various theories. Notwithstanding, Italy would bless him as the real discoverer. At his house in Rome, a visitor may still read the commemorative plaque: "Here resided Andreas Caesalpinus, discoverer of the circulation of blood and the first author of the classification of plants."

BY THE MIDDLE OF THE SEVENTEENTH CENTURY, THE STORM THAT *De motu cordis* had fueled was over in England. Harvey could survey an English scientific world through rose-tinted glasses, very different from the one into which he had introduced his work a quarter century earlier. The battle for Harveyan enlightenment was nearly complete, hardened by victories laboriously won through a life of storm and stress, anxious self-questioning, and severe emotional conflict. But he could never forget the pack of dishonest, inept detractors who were still occasionally harrying him. The decline of Galenism, like the Roman Empire, had resulted from the natural and inevitable effect of immoderate greatness. As time and accident removed the artificial supports, the cracks had begun to open, and the stupendous fabric had yielded to its own weight. Such is the invariable story of many a causal hypothesis that, at its formulation, seems the best possible explanation. But as time passes and new knowledge accrues, it increasingly fails to account for new experimental facts and must be discarded. It was for that reason that Newton excluded "hypothesis" from his field of science, and Poincaré warned scientists to regard hypotheses as mere verbal analogies to be indulged in as warily as possible.

Harvey left behind more than enough keepers of his flame. Everywhere, the younger generation, in their different ways, were turning over the rules and shibboleths that had guided physiology for more than a millennium. The College of Physicians was now dominated by Harvey's supporters. Some, like Jonathan Goddard, warden of Merton College, and Thomas Wharton, physician at St. Thomas' Hospital, were students of his first converts. Others, such as his faithful assistant Charles Scarburgh, became his lifelong friends. In his will, Harvey bequeathed Scarburgh his velvet gown and "all my little silver instruments of surgerie."

Some Oxonian disciples went out into the provinces, whereas others, such as the physician George Bathurst who had attended Oliver Cromwell, stayed on at Trinity in the university town. Bathurst gave a memorable lecture on the nature of respiration in which he presciently spoke of a *spiritus nitrosus* as an inflammable particle in the air that replenished the vital spirits in blood. To augment his income, he joined the medical practice of Thomas Willis and William Petty. Petty, fellow of Brasenose and a founder of the Royal Society, had studied on the Continent and assimilated Descartes's doctrines.

Sir Francis Glisson, now Regius Professor of Physic at Cambridge, and Sir George Ent of Sidney Sussex College, now president of the Royal College, both endorsed the circulation. Ent had met Harvey in

Venice during his Paduan days, and his "Defense on Behalf of the Circulation of the Blood," published in 1641, was the first in-depth affirmation of Harvey's works based on his own anatomical observations. Harvey would bequeath in his will to "Mr. Doctor Ent all the presses and shelves he please to make use of and five pounds to buy him a ring to keep or wear in remembrance of me." But in terms of the heart's function, Ent was thoroughly Cartesian. Glisson supported his pupil, John Wallis of Emmanuel College, to defend a public dissertation on the circulation as evidenced by a letter written by Wallis.

In 1644, the dashing knight-errant Royalist and Harvey's friend Sir Kenelm Digby, a Catholic activist, physician, diplomat, and debonair courtier, passionately endorsed the circulation in his *Two Treatises* published in Paris, in which he effectively refuted Descartes. Digby would become infamous from a scandal involving his prenuptial indiscretions with Venetia Stanley. Indeed, he was so madly enamored of her and strove so hard to safeguard her beauty that he published a book on "Secret Experimentations to Preserve the Beauty of Ladies," which he gifted to Harvey. Alas! The lady died in her prime, and Sir Kenelm was beheaded for his participation in the Gunpowder Plot.

The next decade opened favorably with the appearance of *Corporis humani disquisitio anatomica* in 1651 by Harvey's Oxford colleague Nathaniel Highmore. Highmore was educated at Trinity College and set up practice at Sherborne in Dorsetshire. His massive folio, dedicated to Harvey, described the circulation as caused by a violent pumping action of the heart by which blood was pushed into the vessels. Such a forceful motion, he wrote, not only reliquified the blood grown cold at the periphery but also carried spirits and nutrition. Some of his physiological remarks, however, were soundly medieval. His theory of heart function was Cartesian, derived from Aristotle. Whereas Harvey had transferred the innate heat to the blood, Highmore continued to add fuel to a living flame within the heart chamber.

The following year, Henry Power of Halifax, a correspondent of Sir Thomas Browne, gave Harvey full credit for the discovery in his *Circulatio sanguinis, inventio Harveiana*. Power built an "air pump" and anticipated Boyle in discovering Boyle's law; he also ranked among the original elected members of the Royal Society.

THE LAST WORD HAD BY NO MEANS BEEN SAID IN MIDCENTURY SPAIN, which remained Galenist well beyond the middle of the century. However, unlike his visit earlier, Harvey found a few moderate Galenists who were open to change. Among them was Gaspar Bravo de Sobremonte, an aggressive youth who had served as physician to the Inquisition and was now *physician camera* to the Spanish monarchs Philip IV and Charles II. In his textbook, *Resolutiones medicae*, published in 1649, which would go through several editions, he accepted only those observations that had seemed the most undeniable, and then only as a secondary detail that in no way compromised the general coherence of the Galenic system.[4] Bravo was not only familiar with Harvey but also quoted from Highmore and other contemporaries. He refuted the objections of Primrose and Parisanus.

On the other hand, Mateo Garcia, professor of anatomy at Valencia and a dedicated experimental scientist with much experience in research on eels, frogs, and doves, rebutted the circulation completely. He was unable to duplicate Harvey's findings, or even find the venous valves in humans. Among those who challenged Garcia was the Portuguese physician João Marques Correa, who defended Harvey in disputations at the medical school in Beja, Portugal. He published the most important Portuguese treatise on the circulation in which he surveyed Hippocrates, Galen, Fabricius, Servetus, and Harvey. He summarized the anatomy of the heart and vessels in four chapters along with the "marvelous" movements of the heart and its "strange causes" according to ancient and new doctrines, then outlined a perpetual circulation of blood. He closed by rejecting the arguments that had been raised against Harvey.

The last quarter of the century nurtured a new brand of Spanish physician, contemptuously baptized the "novator," who broke radically with the authority of Galen and embraced the circulation as fundamental to the new medicine. The novators, who gave Harvey his due mead of praise and comment, went public in 1687 with the work of Juan Bautista Juanini, a Milanese physician who had come to Spain as personal physician to Prince Juan José of Austria. In all his books, references to the circulation were numerous and formed an important cornerstone for his own theories and models. Juanini's influence was undoubtedly decisive in making Saragossa the first Spanish university to introduce Harvey's doctrine as a regular part of the syllabus.

The circulation theory was positioned as a foundation not only of a new physiology but also of a new medicine in the most important medical textbook in Spain, that of Juan de Cabriada of Valencia, published in 1687. The essential points of recent discoveries in anatomy, physiology, pathology, and therapeutics were presented. It was inevitable that so peremptory a position should provoke the Galenists. Violent polemics ensued through anonymous public pamphlets, some defending the Valencian's work and others attacking it. The crude recourse of claiming to find elements of a true circulation in ancient medical classics became the traditionalist subterfuge for removing the element of conflict from doctrines whose validity could no longer be denied. And so, Pedro Acquenza y Mossa accepted the circulation but asserted its basic features had been known to Hippocrates and the Greeks.

By the last decade, moderate Spanish Galenists adopted a more constructive attitude. Joan d'Alòs, professor at Barcelona, covered in his 1694 book cardiac anatomy, physiology, blood transfusion, intravenous injections, and the pathologic anatomy of the heart and arteries, discussing at length the origin of blood, the generation of vital spirits, the relation between respiration and circulation, and the nature of the heart's contraction. It remained the most important exposition on cardiovascular function in Spain for decades.

THE FIRST RECORD OF HARVEY'S DOCTRINE IN THE AMERICAN COLONIES appeared in "New England" in 1687 in the philosophical lectures of Charles Morton, pastor of the Church of Charlestown near Harvard College. Morton had been at Oxford at the same time as Harvey and may have met him. He began a manuscript on contemporary scientific thought in England and published it in the American colonies as a *Compendium physicae*. Divided into thirty-one chapters, it was based on the works of Aristotle but incorporated the current ideas of Francis Bacon and Descartes as well as Harvey's experiments. The circulation, with its experimental validation, was spread over nine chapters.

Some years after Morton's death, Benjamin Franklin became interested in the problem of the circulation and explained its mechanics through mechanical glass models. After reading the works of Sanctorius, Harvey, and others, Franklin maintained a lively correspondence on the subject with the influential New York physician Cadwallader Colden. Two other books in the colonies made passing reference to the circulation: an essay on fevers by John Walton (published in 1732), and an essay on health by Grosvenor, which had first been published at London in 1716.

In the Spanish American colonies, Marcos José Salgado, who occupied the chair of medicine at the University of Mexico, published a thesis in 1690 on the anatomy of the heart, as well as a textbook wherein he enunciated the virtues of the circulation with full credit given to Harvey. The heart was described as the main force that drove blood into the arteries; blood that left the heart through the arteries returned through the veins. The pulmonary transit was correctly described. The Galenic idea that the liver was the origin of blood was refuted, but Salgado was vague about the function of the valves in the veins and the capillary circulation, despite being familiar with the publications of Fabricius and Malpighi. On the other hand, Father Jose Ramirez, a naturalist and publisher of several journals and works on natural philosophy, proposed that it had all been really described by Saint Ambrose in a book about Noah's ark! But Harvey, he added, after repeated observations and dissections in animals, had made it clear to a point beyond doubt.

UNLIKE MANY DISCOVERIES IN SCIENCE, HARVEY'S CIRCULATION BE-
came widely accepted in Europe during his own lifetime. It was too
hard to disprove mainly because his argument was based upon careful
direct observation at vivisection and thoughtful experiment. Another
aspect of his reform was the application of quantitative reasoning as
the basis for conclusions about living processes. In some instances, his
doctrine was endorsed less through the power of its truth than by the
theory's compatibility with other dominating interests or natural philosophies, as in the case of Fludd and Descartes. By the end of Harvey's
life, Galen's doctrine had become an archaism, a phantom from the
past. That growing recognition was plainly gratifying. A new course, a
new attitude to the microcosm, had begun.

"CONSIDER THE AIR"

The known is finite, the unknown infinite. Intellectually we stand on an islet in the midst of an illimitable ocean of inexplicability. Our business in every generation is to reclaim a little more land.

—Thomas Henry Huxley

twenty-nine

The Goodness of Airs

Why and how air is requisite for all animals
that breathe as also how air is necessary for
a candle and for fire, I, W H, have seen.

—WILLIAM HARVEY

"CONSIDER A LITTLE MORE CLOSELY THE NATURE OF THE AIR," Harvey had advised when expressing his doubts on the cooling function of air. The study of the circulation led naturally to the problem of breathing. Galen had linked the air breathed in with blood by mixing them in the left heart chamber to form the vital spirit. But if there was no longer a need to appeal to vital spirit, one still wondered why blood mixed with air and what happened when it did. On that matter Harvey offered no precise theory. With the mechanics of a circulation well settled, it was now up to the chemists to investigate what function the circulation could serve in terms of respiration and, in particular, examine the physiological significance of the change in the color of blood in the lungs.

A young crusader named Robert Boyle took up the cause. Chemistry, he held, was the key to the body. It provided the basis for understanding physiological phenomena by illuminating potential mechanisms by which organs received and altered blood. Above all, it could explain *why* blood circulated. Boyle's well-known air-pump experiments still hold a canonical place in science texts and in science pedagogy as a model of how authentic scientific knowledge should be secured. What makes them significant to our story is that they pioneered research on the purpose of the circulation. Working alongside chemists at Oxford, he highlighted a possible particulate (corpuscular) component of air as an explanation for the process of combustion and respiration, setting into motion the quest for such a component that combined with blood to generate life.

Boyle, the Anglo-Irish "Skeptical Chymist," as he was called, was a younger contemporary of Isaac Newton and "the chief wonder of the English scientific world" after Newton. The fourteenth child of the second Earl of Cork, a political *nouveau riche* of the early Stuart period, he was born at Waterford, Ireland, a year before Harvey published *De motu cordis*. Educated at Eton and then in France, Switzerland, and Italy, he became familiar with the work of the recently deceased Galileo.[1] Tall, pale, and emaciated, he showed poor health, weak sight, and a bad memory. Suffering great pain from a kidney disorder, he was professionally driven by the desire "to relieve some languishing patients the more easily from their sickness; for certainly, our common remedies are very ineffectual." Mercifully, his own death "made an end of him with so little uneasiness, that it was plain his light went out merely for want of oil to maintain the flame." He was buried in St. Martin-in-the-Fields in London. The old church was pulled down in 1721, and his remains were lost. His manuscripts and notes on chemistry were entrusted to the Enlightenment thinker John Locke, who was himself an able physician and chemist and editor of Boyle's *General History of the Air*.

Boyle had decided early in life to devote himself to science and spent about six years as an apprentice to the foremost scholars of the day. His life was a contrast to the life of Newton, who, from humble birth and little fortune, had risen to eminence by sheer ability and had used his scientific genius to secure wealth and position. By the age of thirty-four, Boyle had assured his own reputation as the creator of a new experimental chemistry and, as the "Skeptic," he had discredited the four Empedoclean elements and the three *primae materiae* of the Paracelsians. After a decade's sojourn at Dorset, where he engaged in anatomical dissection, he settled at Oxford in 1654, taking up lodgings next to University College on High Street soon after Harvey had left to end his days at London.

Renowned in history as a scientist and chemist, Boyle was no less notable in the eyes of his Oxford contemporaries as a Christian gentleman who centered his whole life on the Christian practice. Like Newton, religion rather than science was the foundation of his being. Throughout his life, he spent about 1,000 pounds annually in works of private charity and on Christian missions and the propagation of the Gospel; he had the Bible translated into Irish, Welsh, Hindi, Turkish, and Malayan. It is likely that Boyle's first work on moral essays and his early experiments may have given Jonathan Swift the idea for his *Gulliver's Travels*. More, perhaps, than any of his peers, he sought and found the power, wisdom, and the hand of the Christian God in all that he described. Among his papers at the Royal Society were numerous fragments on miracles. The Bible had reported miracles, and he neither could, nor would, depart from his belief in them. But the order of nature, and the unfailing rule of natural law over matter, dominated Boyle's imagination as no miracle could. The Newtonian conception of a universe ruled by eternal immutable laws was one of the aspects of nature that excited his religious awe the most, however much it may have been an unconscious presupposition.

OXFORD IN THE SEVENTEENTH CENTURY WAS THE MOST EXCITING SCIentific community in England. The university was expanding, not merely with stone and mortar but with students and scholarships, as both merchant and gentry were discovering that a college education was desirable for personal improvement as well as public advancement. Lands were bestowed to enhance the foundation's revenues, library facilities were improved, and a book trade settled within the campus. Richard Davis, a bookseller, maintained a stock of well over thirty thousand volumes and, in shops like his, scholars gathered to socialize, examine folios from a dozen continental publishing centers, and see their own writings through the local press. In addition to the Regius chair in medicine, there were Savilian professorships in astronomy and geometry and a Sedleian endowment to support a lecturer in natural philosophy. A "Physic Garden" was organized under the supervision of Jacob Bobart, a German botanist. The Tomlins Readership in anatomy provided grants for annual dissections and lectures in anatomy, and a room by the Quadrangle was refurbished to be an anatomical museum with dissection facilities.

Unlike our own times, the written word was scarce and expensive; libraries were few, and the Bodleian was in its infancy. Coffeehouses and ale taverns were the common centers for groups to meet according to their interests. It was at a tavern that the famous discussion between Wren, Halley, and Hooke on the problem of gravitation took place and sowed a seed leading to Newton's *Principia*. The early meetings of the Royal Society were invariably held at an alehouse in the Strand, where new explanations on the physiology of blood, heat, and respiration were debated and developed on the basis of novel atomistic and mechanical philosophies that were emerging from Europe. Because they had no fixed place of meeting, the members called themselves "the Invisible College."

Good fortune blessed them from an unexpected quarter when Charles II saw fit to dabble in science. Experimentation grew fashionable at court and the Invisible College, under the king's favor, became the Royal Society on November 28, 1660. It was founded for the "Improving of Natural Knowledge," and its motto, shown on the coat of arms, declared *Nullius in verba* (Take no one's word for it). It was an

expression by the fellows to withstand any sort of domination by authority and to verify all statements by an appeal only to facts determined by
experiment. It also promised to cultivate high standards among physicians after the model of Thomas More's *Utopia*, which had displayed
great sensitivity to the public health needs of urban communities. There
were no women among its fellows and wouldn't be until 1945.

MOST OF BOYLE'S SIGNIFICANT SCIENTIFIC WORK WAS CONDUCTED
and published between 1660 and 1673. Inspired by the 1654 publication of Otto von Guericke's famous demonstration of the air pump at
Magdeburg, and encouraged by his youthful and inexhaustible assistant Robert Hooke, Boyle had Hooke build a "pneumatick engine"
(an air pump) that could create a vacuum chamber to investigate the
physical properties of the air and the effects of an airless environment
on larks, sparrows, mice, ducks, cats, and cheese mites. He noted that
the volume occupied by air was inversely proportional to the pressure applied, an observation that would eventually become known as
Boyle's law. He assumed that air consisted of corpuscles defined in
the manner of René Descartes, and in that notion he anticipated the
kinetic theory of gases. Among his many achievements as a chemist
were the preparation of phosphorus, the collection of hydrogen in a
vessel over water (he called it "air generated de novo"), and the study
of crystals as a guide to chemical structure. Above all, he demanded
that chemistry should be founded upon a substantial body of experimental observations, and he called in particular for the quantitative
study of chemical changes.

In a series of astonishing experiments the likes of which had never
been attempted before, Boyle affirmed that both animal life and a flame
needed atmospheric air. More than a millennium earlier, the Alexandrian Erasistratus had thought about metabolic processes by confining
a fowl in a jar and making measurements of the bird's weight and that of
its food and excreta before and after digestion. But there was no record

of "respiration" or heat exchange. Candle flame and animals alike, when enclosed by Boyle in vessels, absorbed part of the air available to them. He suspected that this life-sustaining component could be a vital quintessence that was mixed with the atmospheric air. The analogy between life and combustion, which had been distinctly recognized by Aristotle and Galen, was not lost either. Clearly, a lit candle, too, depended upon some odd substance in the air that, "enabling it to keep flame alive does yet by being consumed or depraved, render[s] the air unfit to make flame subsist." The life of an animal left in a milieu from which air had been partially pumped was more swiftly shortened than if all the air had remained. Even a highly combustible substance like sulfur could no longer be inflammable in a vacuum.

From such experiments, Boyle became convinced that "the necessity of fresh air to the life of hot animals suggests a great suspicion of some vital substance, if I may so call it, diffused through the air, whether it be a volatile nitre, or rather some anonimous substance, syderial or subterranean, but not improbably kin to that which I lately noted to be so necessary to the maintenance of other flames." He continued, "Wherefore I have sometimes been inclined to favorable thoughts that air [is] necessary to ventilate and cherish the vital flame, which they do suppose to be continually in the heart. For we see, that in our engine [Hooke's vacuum pump] the flame of a lamp will last almost as little after the suction of the air, as the life of an animal." He put the excised heart of an eel in a container, evacuated the air with his pump, and observed that, although the heart had "bubbled and grew tumid," it continued to maintain a regular beat. He satisfied himself that a heart may beat without blood and, for brief intervals, did not need air.

Air was present in water too. Fish did not survive when introduced to water that had been boiled vigorously to remove all its air. "It had been noted by divers authors," Boyle remarked, "that fishes soon die in ponds and glasses quite filled with water, if one be so frozen over and the other so closely stopped that the fish cannot enjoy the benefit of air."

EVER SINCE THE GREEKS, AIR WAS CONSIDERED ONE SUBSTANCE, AND differences in reaction were attributed to variances in purity. Now Boyle took a vital first step by calling atmospheric air not an element, as was hitherto believed, but a compound, "for this is not, as many imagine, a simple and elementary body, but a confused aggregate of effluviums from such differing bodies . . . perhaps there is scarce a more heterogeneous body in the world."

In both respiration and combustion, an interaction of some sort seemed to occur between a special substance in the air and a special substance in the body. But he did not conclude that the substance required for the dance of candle flames and the life-giving flame in animals was one and the same. Nor did he believe that the substance was necessarily a constituent part of the air itself but believed it could be some "anonimous substance, syderial or subterranean." Regrettably, though he connected the phenomena of respiration and combustion, he held the function of air to be quite different in the two. In respiration, it served only to carry away noxious "steams" generated in the body. But combustion resulted from special "form[s] of corpuscles" existing in the air.

Chemistry had still a long, painful, and convoluted path to traverse before it could shake itself clear of relying on special effluvia, or "spirits," for every action.

With little more than a suspicion that respired air did something other than just cool the heart and receive the "waste" from the lungs, Boyle set aside his studies on respiration. Respiration for him, at that moment, had become more, rather than less, mysterious than when he had begun his experiments. But he *had* debunked the concept that the key to the problem of respiration and combustion was to be found in the physical properties of the air, such as its temperature or elasticity. It consisted, instead, in something being added to air or something being subtracted from it that was necessary for their support. That was the reason why an animal died, or a flame was extinguished, when kept in a limited amount of unrenewed air. He had come close to a knowledge of, and yet missed, the true function of air in respiration and combustion in Harvey's circulation.

If Boyle's conclusions looked rather opaque, Robert Hooke, his precocious assistant, set out to illuminate them. He continued the experiments during the winter of 1662.

As a child growing up on the Isle of Wight, Hooke was a sickly lad born with a skeletal defect that gave him a twisted frame. He was not expected to survive infancy, but he grew up to be studious and inventive. At thirteen, he was apprenticed to the painter Peter Lely, but left him to register as a pupil at Westminster School. There, within a week's time, he had mastered the first six books of Euclid and later "learned to play twenty lessons on the organ" and had invented "thirty several ways of flying." At eighteen, he entered Christ Church, Oxford, at about the same time as Boyle.

Hooke and Boyle complemented each other admirably. Hooke had a quick though impatient mind, and his taste and genius, coupled with a marvelous mechanical skill, directed him to mathematical problems: he invented the balance spring that would revolutionize watchmaking and determined that the stretch of the spring, or any elastic body, was proportional to the force applied, which became known as Hooke's law. He made microscopes of astonishing quality and built an excellent vacuum pump, though Thomas Hobbes notoriously disparaged it publicly by claiming that it did not create a perfect vacuum and hence all Boyle's research conducted with it was suspect.[2] Boyle, on the other hand, was thorough and cautious, and though skilled in chemical operations, had less interest in mechanical devices.

At the age of twenty-seven, Hooke became Curator of Experiments at the newly formed Royal Society. It was a difficult job. The members—gentlemen, lordships, and an occasional bishop—sat in opulent chairs and discussed one subject after another, often aided and abetted by a selectively invited, vigorously partisan, educated public. Questions arose that could be settled only by experiment, and one of the young curator's duties was to design the required experiment and perform it before the society at the next meeting. If none were assigned, he was expected to invent some "three or four considerable experiments" to keep the "Fellows' minds stretched week after week in every aspect of the new experimental, mathematical philosophy," which was the whole purpose of their association. As William Petty, one of the members, put it, Hooke's was indeed a mind of "tumultuous versatility."

In his first four years, at least ninety experiments were performed on live animals. It is no exaggeration that Hooke, over his forty-year tenure, took upon himself, more than anyone else, to singlehandedly resolve the many difficulties faced by the nascent society and contributed most to its success. His vitality and energy were prodigious, and his capacity for work was legendary. Urged by Christopher Wren, he published one of the first books, *Micrographia*, of drawings of microscopic animals that showed how the world looked when viewed through a microscope, and he was the only man of his century who practiced the new art of "auscultation": he listened to the voice of the heart and lungs by applying his ear to the patient's chest.

THOSE WHO CAME TO SEE HOOKE, DRAWN BY HIS FAME, MUST HAVE been surprised at his appearance. A short, crooked figure with shrunken limbs and uncombed hair falling like a shaggy mane over an ashen countenance, he appeared to grow even smaller and more deformed with the years.[3] The vicissitudes of fate, it seemed, must have passed over his head for he held it, plaintively, rather on one side. His character was as unprepossessing as his body: a continuous suffering from headaches, dizziness, inflammations, insomnia and nightmares, vomiting and indigestion (probably worms) had made him ill-natured and a hypochondriac. Vain and jealous, he was despised by many contemporaries.

But behind those unfriendly sunken eyes and disheveled locks burned the fire of genius. The very fact that he served, in turn, as assistant to "that prodigious young scholar" Christopher Wren (to borrow John Evelyn's phrase), Robert Boyle, and Isaac Newton established his greatness as well as his shortcomings. Hooke was not content to make only half the discoveries of his age; being more sensitive than most of his colleagues regarding the increasing importance of priority in scientific discoveries, he also claimed the other half! When Newton's *Principia* was in publication, Hooke was so insistent that part of the work had been stolen from him that Newton, furious at the charge of plagiarism, virtually suppressed a third of the volume. He acrimoniously resigned from the Royal Society (an unheard act that, fortunately, was not accepted) and swore that as long as Hooke remained its curator, he would publish no further scientific work. Newton was as good as his oath. He abandoned his scientific studies and buried himself in biblical and Hermetic ones. Later, when Sir Isaac completed his *Optics*, he refused to publish it for almost a year after the bitter tongue of Hooke had been silenced forever. In 1703, when Hooke finally died, Newton accepted the position of president of the Royal Society. His first order was the burning of Hooke's portrait. And having laid Hooke triumphantly to rest, Newton turned to an all-out feud with an upstart German philosopher named Gottfried Wilhelm Leibniz, who was claiming credit for discovering the calculus, Newton's pet invention.

HOOKE KEPT A METICULOUS DIARY IN WHICH HE RECORDED HIS EXperimental measurements as well as notes on the weather, his rare and precious restful nights, his personal medical ailments, and the workings of his body, especially his bowels as subjected to his regimes of purgatives and emetics, and even his orgasms. He never married, though he was a daily frequenter of taverns and coffeehouses: his diary mentions an astonishing 154 such places that he patronized. His sexual relationships were invariably with in-house maidservants as well as his brother's teenage daughter Grace Hooke, who died prematurely.

His final years were marred by uncontrollable rages, paranoid vindictiveness, and occasional mental instability, darkened by disease, blindness, and disappointment. In death, as in life, he remained an unattractive sight. The "lice were so thick" on his emaciated corpse that "there was no coming near him." But when he passed away at sixty-eight, virtually the entire membership of the Royal Society attended his funeral to mourn. When an iron chest in his chamber was pried open, it was found to hold nearly 800 pounds in cash and another 300 in gold and silver (almost 1 million pounds in today's values!). His fine library of three thousand books was sold at auction for more than 200 pounds. He died as he had lived—a twisted old miser.

WHEN GEORGE ENT MET SAMUEL PEPYS OVER ALE AT THE CROWNE Taverne, he cautioned that it was "not to this day known or concluded on among physicians" how respiration "is managed by nature or for what use it is." Fresh lines of battle were clearly needed. When the Royal Society reconvened in early October, Hooke returned to the study of respiration with a more rewarding series of experiments. His first trials, and the impressions of his combustion theory, were summarized by Thomas Birch in *The History of the Royal Society*.

The new demonstrations, which were held at Gresham College, were witnessed by the stocky, testy German Henry Oldenburg, the Royal Society's secretary, who sometimes wrote his name as "Grubendol." Pepys, too, was there on his first visit to the society as a member and recorded in his diary that the whole experience had been "a most acceptable thing."

Hooke demonstrated conclusively that the significance of breathing was to supply fresh air to the body.[4] He opened the thorax of a dog and removed the diaphragm and then kept the lungs supplied with fresh air "by means of a bellows and a certain pipe thrust into the windpipe of the creature," a process somewhat resembling the modern procedure of endotracheal intubation. The animal was kept alive by such artificial respiration in the absence of all movements of the chest wall, proving quite convincingly that the essential business of respiration was carried out only in the lungs (a similar experiment had been described over a century earlier by Vesalius). The movements of the chest were useful only so far as they brought about the expansion and collapse of the lungs.

Another demonstration was conducted with the Cornishman Richard Lower, Hooke's friend and pupil. An intubated animal was kept alive equally well even without any movement of the lungs by distending the motionless lungs with small powerful blasts of fresh air using the bellows. The driven air escaped continuously through minute holes previously pricked in the lung by a very sharply pointed penknife. The findings proved that the motions of the chest *as well as the lungs* were merely an incidental feature of respiration. The vital qualities were imparted by the exposure of blood in the lungs to fresh air and not by any organ movement. In the same experiment on the artificially ventilated animal, Hooke demonstrated that the heart continued to beat so long as the motionless lungs were supplied by puffs of fresh air.

The whole event was stunningly effective, albeit one that Evelyn noted in his diary "of more cruelty than pleased me." Clearly, it was fresh air, and not any heat within the heart, that imparted a vital spirit to blood. From several additional trials, it seemed probable that the heart could continue to beat ceaselessly so long as the artificial ventilation was maintained. Hooke concluded correctly that the primary function of breathing was the admixture of air with blood and that "some mixture of air with the blood in the lungs might give that floridness" to blood. That was a groundbreaking demonstration of the "purpose" of the pulmonary circulation. As further proof, dark venous blood, when shaken with air, became bright red.

In 1665, in the year he published his *Micrographia*, Hooke proposed a first step to a new theory of combustion. Air, he speculated, consisted of different components that differed in their *chemical* behavior. One of those parts was active in combustion, and the same part must be operational during respiration. The active constituent, he suggested, was similar, though not identical, to something that was contained in "nitre."

AMONG CHEMICAL SUBSTANCES, "NITRE" (POTASSIUM NITRATE; SALTpeter) had for long been known for its fertilizing and explosive properties. Since lightning and thunder also possessed the same volatile properties, some sort of niter was assumed to exist in the ambient air. By decomposing what he had claimed to be niter, a Dutch chemist named Cornelius Drebbel had prepared a gas capable of maintaining life. Paracelsus, too, had argued that air was a form of nourishment absorbed by the lungs, just as food was absorbed by the stomach. There must be a kind of refreshment necessary for life in the air, Hooke speculated, "a nitrous quality" that, when spent or entangled, made the air unfit. Boyle concurred. Animals died quickly without a continuous supply of fresh air and did so almost immediately when placed in a vacuum. Hooke was clearly probing his way with great difficulty toward a single constituent of air that was functional in combustion as well as respiration.

In the fall of 1669, Boyle performed another series of experiments before the Royal Society. This time he enclosed a mouse in a sealed vessel together with a mercury pressure gauge. After two hours, during which time the mouse continued to breathe, there was no apparent change in the air pressure. Boyle concluded that air which had become unfit for respiration continued to retain its "spring" or "wonted pressure." Such observations, which marked the beginnings of respiratory physiology, led Boyle to concede that whatever it was that was vital for combustion and respiration had to be an imperceptibly small bulk of the air since the volume of air was not diminished. In another experiment, Boyle took a lark and then a hen sparrow and sealed them with a burning candle under a glass receiver. Every time the investigation was repeated, the bird outlived the flame. It seemed that the death of birds "proceeded rather from the want of Air, than that the Air was over-clogg'd by the steams of their Bodies."

Such observations led him to ask whether the common flame of the candle and the "vital flame" within the bird were maintained by distinct substances in the air or whether a single constituent of air nourished both but was used up more rapidly by the common flame while there still remained enough for the more temperate vital flame within the bird to keep it alive. Earlier, John Evelyn recorded in his diary that he had witnessed "divers Experiments in Mr. *Boyls Pneumatic* Engine" in which "a chick was killed outright by evacuation, but a snake was merely sickened."

During the ensuing decade, Hooke turned episodically to combustion, respiration, and the life-air problem. One of those experiments involved enclosing *himself* with a candle in an airtight cask for over fifteen minutes during which time the candle went out, and then his ears began to "pop" uncomfortably. In 1672, he informed the society that, during the process of breathing, something in air essential to life was dissolved in blood and something that was unfit for life was discharged back into the air. Death resulted from a satiating of the dissolving part of the air, thus making the remaining part effete and useless for maintaining life.

He reiterated that a nitrous component in air might properly be regarded its "vital" part, and it was that portion that supplied the "vital spirit" to burn in flaming bodies and maintained the life heat and motion of animals. Although Hooke's theory of combustion did not gain wide acceptance, the suggestion that atmospheric air contained some "nitrous" quality that rendered it fit for the support of respiration and combustion seemed to be plausible. It would be destined for a long history in the pursuit of a "purpose" for Harvey's circulation.

thirty

Oxford Chemists

Experience is not at fault; it is only our judgment
that is in error in promising itself from experience
things that are not within her power.

—LEONARDO DA VINCI

WHEN BOYLE WAS AT OXFORD, HE BECAME THE TEACHER AND
adviser of Richard Lower, a young man of substantial wealth who
had grown up at his ancestral home (which still stands) at Tremeer, in
Cornwall.[1] The story of Richard Lower begins and ends in Cornwall.[2]
There he was born and there he lies buried. There, too, he married the
wealthy widow next door, and he sought there a quiet retreat when the
court at London turned against him. Like many others, he could never
quite forget the rugged moors and stormy sea of his birthplace, and
neither Oxford nor London could ever detract his affection for his own
West Country.

After finishing at Westminster School at age seventeen in the year of
King Charles's execution, Lower was elected to a studentship at Christ

Church at Oxford, where he befriended Hooke. Before long, he joined the well-known physician Thomas Willis as a research assistant, a position he held for ten years. Their collaboration was, without doubt, mutually beneficial. Willis was constantly suggesting research problems to Lower to see whether his own shrewd ideas could stand the test of trial. Lower's skill in experimental design and his precise knowledge of anatomy were no less valuable to Willis. Following Willis to London soon after the Great Plague, Lower outlived his senior associate by sixteen years and inherited his medical practice to become the most illustrious doctor at Westminster and in London, as well as court physician to Charles II, whom he attended during his final illness.

THE FIRST EVIDENCE OF LOWER'S INTEREST IN AIR AND BLOOD appeared in a letter he wrote to Boyle on June 24, 1664, in which he proposed his intention to study "the reason for the different color of the blood, the one being florid and purple-red and the other dark and blackish." The following year he graduated as Bachelor and then Doctor of Physick. His career began on the wrong foot when he published a tract *Diatribae de febribus* in which he erroneously stated that venous and arterial blood in the lungs were the same dark color and that the florid tinge of arterial blood was acquired by passage through the left auricle of the heart, where it was mixed and kindled! Harvey, too, had insisted that arterial blood was essentially the same as venous blood despite the striking difference in color; he had foreseen an obvious obstacle to his circulation: If there was only one system, how could there be two different kinds of blood in it? Lower corrected his ghastly error in a later volume, *Tractatus de corde* (*A treatise on the heart*) that he published in London early in 1669, three years after he moved there.

De corde was a book of discoveries.[3] Lower was a dedicated and avid anatomist. The investigations and conclusions reported had a significance beyond the spare terms in which they were presented. It had its origins in the summer of 1667, ten years after Harvey's death, when

Boyle introduced Lower to the Royal Society. It claimed to complete the edifice that Harvey had begun and to address those matters Harvey had promised to explore: the structure of the heart muscle, the quantity of blood in the vascular system and the velocity of blood flow, and the effects of aeration on blood in the lungs. Indeed, Lower was called Harvey's successor and a leading exponent of the anatomy and physiology of the circulation. An Oxford antiquarian, Anthony Wood, pointed out that Lower often skipped mass in favor of dissection in his rooms adjacent to Christ Church College.

Lower was an anatomist by training but, by inclination, an experimental physiologist, and his work, in its crisp and direct structure and content, reflected both traits. His experimental designs, during his hectic activity at the Royal Society in the autumn and winter of 1667–1668, were precise and rigorous, and his demonstrations simple and convincing. Also marked were his honesty regarding his previous error on the color of blood and his candid expressions of indebtedness to fellow investigators. Through a series of shrewd experiments and logical arguments, he completed the dismantling of the Galenic system that Harvey had begun. Yet he did not propose a general physiology that would encompass all his individual findings.

Like Boyle, Lower believed in, and found evidence for, a divine plan. Every part of the heart was not merely workable but perfect, the masterpiece of a supreme intelligence. He marveled that the left ventricle, which drove blood through the whole body, was created more muscular than the right one, which forced blood only through the lungs. Valves in the heart and in the veins were devised so that they permitted flow in one direction but prohibited the reverse. At the exit from the left ventricle, blood was thrown straight up; but the "divine Artificer" had foreseen a potential problem and had provided the construction of an aortic "arch" that directed the main flow of blood downward to the body. Even a thing as arbitrary as the network of arteries and veins turned out, upon close inspection, to be a product of divine wisdom and design.

LOWER EXPOSED AND IMMOBILIZED THE LUNGS AT VIVISECTION, VEN-tilated them artificially, and collected, from a pulmonary vein, the blood that had not yet reached the heart. He found, as Colombo had, that it had already acquired its vivid color. If the flow of air to the lungs was arrested, the blood in the vein turned black, but when air was re-admitted, it became crimson again. When he drove dark venous blood through the artificially ventilated lungs of a dog newly dead, the blood discharged into a dish from a severed pulmonary vein was bright red, as if it was withdrawn from an artery in a living animal. "On this ac-count," he concluded, "it is extremely probable that the blood takes in air in its course through the lungs and owes its bright color entirely to the admixture of air." The proper nature and function of respiration were to supply a life-giving quality, a "vital spirit," from the air to blood. Servetus's theology had metamorphosed into Lower's physiology! Here, finally, was the true purpose of Harvey's circulation!

Lower took great pains to deliver the coup de grâce to the ebulli-tion theory of Aristotle and Descartes. Descartes had insisted that there was a ferment within the blood in the left ventricle. It was troubling for Lower to imagine how a ferment of any nature could be supplied continuously. Replacing a dog's blood with a *noneffervescing* fluid that could not support fermentation produced the same heart contractions (thus refuting the need for fermentation) so long as the animal was kept alive by artificial ventilation. Even more telling were Lower's logical experiments and inquiries. The active phase of the heart's action was its contraction, like a pump, as Harvey had shown and Lower now verified. Yet the ebullition theory required that ejection occur during the heart's dilation, like Aristotle's analogy of boiling milk overflowing. And what sort of ebullition could be so regular as the heartbeat? Why did that effervescence not take place in the auricles, which were the first parts of the heart to live and the last to die? In any case, the left chamber was not large enough to permit exuberant effervescence. Finally, Lower confirmed that blood did not enter the heart drop by drop, as Descartes had imagined, but in a large stream as Harvey had demonstrated.

Lower agreed with his contemporaries that the function of blood was to maintain the heart's temperature but dispossessed that organ of its vestigial role as the center of animal heat. When he inserted a finger into the heart chambers during vivisection, he did not feel heat of any greater intensity than the rest of the body. The fat that lay directly on the surface of the heart, he reasoned, would not remain solidified the way it did if it was exposed to such high temperatures as were proposed by the fermentation theorists. He was not at all certain that animal heat was produced only in the heart, or that blood was warmed only by it, because other muscles generated heat, as well, by their activity. Thus, the movement of blood, he concluded, was independent of any heating. In an appendix attached posthumously to his colleague John Brown's *Myographia nova*, published in 1697, he disclosed that when blood was allowed to stand in a container, it separated into a watery matter, which he called *succus nutritius*, and a dark reddish matter, which was the blood itself. It was the latter red content (the red blood cells), he believed, that contained the vital substance.

Surpassing Harvey, Lower proved beyond doubt from observations on boiled heart specimens that the organ was a muscle equipped with two main layers of spiral muscle fibers. Blood moved only by the heart's muscular contraction carried out by those fibers. There was no option but to consider the heart simply as a pump as Harvey had always maintained. With Newton's new and more exact physics now at hand, Lower extended Harvey's quantitation by computing the output of the ventricles and the velocity of blood in the arteries of animals. Accurate measurement of the blood pressure in humans had not yet been attempted.

Shortly after the accession of the Catholic James II, Lower was deprived of his position at court. The king relentlessly punished all who opposed his religion and even tried to impose his authority at Oxford, where the fellows accepted expulsion rather than James's Catholic candidate as their president. His practice had been dwindling since the collapse of the Whig Party, and Lower now spent more time at home in Cornwall. There he continued to revise his magisterial *De corde* for successive editions. He did not live long enough to enjoy the more tolerant reign of William and Mary. He died in his house on King Street, Covent Garden, from a chill he had caught a few nights earlier while helping to put out a chimney fire. His body was taken back to Cornwall, and he was buried where he had been baptized fifty-nine years earlier, in the parish church of St. Tudy.

thirty-one

Niter, Niter, Everywhere

There should be no doubt at all now that an aerial something absolutely necessary to life enters the blood of animals by means of respiration.

—JOHN MAYOW

T HE CORNISHMAN WHO BRILLIANTLY SYNTHESIZED THE FIND-ings of Boyle, Hooke, and Lower was John Mayow, also one of Boyle's pupils. He published his entire life's work in 1674 as a single manuscript, his *Tractatus quinque* (*Five medico-physical treatises*). Born a year after Newton in the parish of Saint Dunstan-in-the-West in London, he grew up to be a precocious and somewhat unprincipled young man. He was remarkably adept at grasping and absorbing the ideas and observations of others whose experiments he had been privileged to witness. He was equally eager to make them his own, and he was more than reticent in acknowledging their true sources! But Mayow had an insight into physiological theory that his contemporaries did not. He saw how respiration, heat, the circulation, and the heart's muscular motion could be yoked into a general framework of respiration and combustion. Devoted as he was to science, he took a degree in law and became noted for his legal practice, especially in the "summertime at Bath." He eventually migrated to London and, in 1678, was admitted a fellow of the Royal Society. The following year Mayow was found dead at Covent Garden at the all-too-early age of thirty-seven, having married just a little earlier though not altogether to his satisfaction.

In 1670, Boyle had found no reduction of air during respiration. In 1673, Hooke had reported to the Royal Society that, after multiple failures, he had found that the volume of air was indeed diminished by one-twentieth during combustion. That finding remained unverified, and the issue attracted little further serious attention at the society until the publication of Mayow's *Tractatus*.

Rather than working with a vacuum pump or a sealed vessel within which a mercury barometer had been enclosed, as Boyle had done, Mayow utilized a sealed glass vessel inverted over water. In that way, he observed changes in the volume of enclosed air caused by respiration or a flame by measuring the rise in water level as the animal breathed or a candle burned. Boyle's apparatus had concealed reductions in the volume of air that could have taken place. Mayow's experimentation over water revealed them. "I have ascertained from experiments with various animals that the air is reduced in volume by about one-fourteenth by the breathing of animals," he wrote.[1] A candle, too, became extinguished when a fourteenth part of the air had been consumed. He confirmed that "an animal enclosed in a glass vessel along with a lamp will not breathe much longer than half the time it would otherwise have lived" when it was placed in the same chamber without a candle. He could now infer from his quantitative assessments that air was deprived by breathing in very much the same way and amount as the burning of a candle flame, and that "we must believe that animals and fire draw particles of the *same* kind from the air" (italics added). More striking, he inferred that "it clearly appears that animals exhaust the air of certain vital particles . . . that some constituent of the air absolutely necessary to life enters the blood in the act of breathing."

Those experiments validated, as Boyle had suspected, that the whole air was not necessary for respiration and combustion but only a small part of it. Air was hence not a single homogenous entity but a mixture, as Boyle had proposed, for it was not air but rather a single specific constituent among several that was removed. That constituent Mayow named the "nitro-aerial" component. He was confident that he had found conceptual and experimental verification for the analogy between combustion and respiration and could now assert just what Boyle had hesitated to conclude.

MAYOW OFFERED AN EXPLANATION RESEMBLING HOOKE'S AND shared with it the same difficulties. The four elements of Greek natural philosophy (air, water, fire, and earth) had been modified in medieval times only insofar as "earth," in its different forms, was held to consist of three other elemental substances: salt, mercury, and sulfur. All combustible substances were considered "salino-sulfurus" combinations of salt and sulfur particles that were thrust out when they were heated and then became visible as flame. Their specific active ingredient was niter (sodium and potassium nitrates), upon which the chemist Paracelsus had speculated, and which we now know to be rich in oxygen and a powerful oxidizing agent.

Mayow invented a mythic five-element theory, the antecedents of which were alchemical. The two active elements were sulfur and another subtle, agile, and ethereal "nitro-aerial spirit" that he proposed as the counterpart of mercury. The three passive elements (salt, water, and earth) were called the *terra damnata*. The "nitro-aerial spirit" (*spiritus nitro-aereus*) was the active principle, the new "vital spirit." He then speculated the nitro-aerial particles to be everywhere: from the sun's rays and the lightning from clouds to iron filings that sparked when struck. He did caution that "although flame and life are sustained by the same particles it is not on that account to be supposed that the mass of the blood is really on fire."

The idea of a "nitrous spirit" had been floating in the Oxford air since as early as 1654, but it was Mayow who unambiguously asserted that the nitro-aerial spirit was what made air intrinsic to respiration and combustion. An inability to isolate and identify the nitro-aerial particles left his theory unproven. Techniques for the isolation of "airs" were still in the air.

BOYLE HAD MADE A GREAT DISCOVERY IN PROVING THAT, BY CALCINA-tion or oxidation, metals gained weight. A body *gained* weight when burned, and Mayow inferred that the nitro-aerial spirit must have com-bined with the burning substance. Such particles in blood, formed by the union of sulfur in the air with the salt in blood, caused "a very marked fermentation" and with it the effervescence responsible for an-imal heat. That was a major shift from the ancient doctrine of an unex-plained "innate" heat. Like Van Helmont and Descartes, and contrary to Harvey and Lower, Mayow held fermentation to be the source of body heat and the key to all physiological processes. "We do not need to have recourse to an imaginary Vital Flame that by its continual burning warms the mass of blood," he wrote.

Respiration involved an agitation of the particles in air. Combin-ing his theory with Lower's observations that venous blood changed color on exposure to air in the lungs, he postulated that nitro-aerial particles in air united with the saline-sulfurous particles within blood during their passage through that organ. That was Mayow's historical contribution, and it set the stage for the discovery of oxygen. Descartes's scheme, in which the heartbeat resulted from a process of heating and the consequent rarifying of blood in the heart chamber, was specifically criticized.

Supporting Harvey and Lower, Mayow concurred that the heart contracted because it was a muscle. Owing to its continuous labor, it caused the active mixing of nitro-aerial and sulfurous particles in its chamber, producing heat. Heat accompanied all muscular activity or violent motion and arose from an effervescence of specific corpuscles in air and blood—a theory of friction that had been mentioned even by the great Newton in his *Opticks*. It marked a change in physiological orientation by blending chemistry with physics.

THE SIGNIFICANCE OF THE OXFORD CHEMISTS, AS BOYLE, HOOKE, Lower, and Mayow were called, was their major advance in demonstrating a "purpose" of Harvey's circulation. Together, they concluded that air played a vital role in combustion and respiration, with part of the air being absorbed in these processes. They highlighted a particulate (corpuscular) explanation for the process of combustion and respiration based upon the framework that Boyle had set up in his well-known law of gases. Boyle was even receptive to the idea that special patterns of corpuscles were the stuff of life. Mayow attributed all phenomena to the operations of his favorite invention: the nitro-aerial particles. He, and others, made it the basis of speculations when factual evidence was not evident. It was left to the pioneers of the following century to *isolate*, for the first time, the components of air that the Oxford Chemists had only hypothesized.

The Oxford group was not destined to continue for long. Old friends died or left town. Timothy Clarke passed away in 1668 and Peter Stahl, a chemist of the Oxford Club, crossed the bar sometime in early 1670. Ralph Bathurst, who was Harvey's supporter and friend, reverted to a church career not long after the Restoration in 1660. Willis was given a scandalously extravagant funeral at Westminster Abbey in 1675. Mayow, himself, died suddenly in 1679. Wren and Locke plunged into different distractions: the former became almost totally immersed in the task of rebuilding London after the great fire of 1666, and the latter moved away from science and medicine (he was an accomplished physician) into political theory. Nathaniel Highmore lived the rest of his life in the provinces. That left, besides Hooke, only Robert Boyle, who had been so prominent a catalyst. Boyle withdrew completely from the

Royal Society in 1674. He died at his home in Pall Mall on December 30, 1691.

With the Oxford Chemists, the work of the English school of medical chemists came to an end. The tradition of the school was continued to some degree by the Reverend polymath Stephen Hales (more about him later), but it was eclipsed in the subsequent period when a fresh wave of chemistry, actually a retreat into alchemy, spread over Europe from Germany in the form of the weird invention of phlogiston and the phlogiston theory.

thirty-two

Phlogiston

Interpretation is guesswork of varying degrees of
sophistication, one false guess that fits the facts,
and it is only too easy thereafter to fit other "facts"
to the guess.

—CARL WILHELM SCHEELE

THE NEXT ADVANCE IN RESPIRATION PHYSIOLOGY BEGAN IN A
country where civilization still had the energy of newness. Scotland!
"Auld Reekie," as Edinburgh was affectionately called, became a home
of brilliant talk and genial company. Sir Walter Scott furnished the romantic imagination for more than a century, James Boswell published
one of the most permanently entertaining biographies in the English
language, and Robert Burns composed the first great popular love lyrics
in the vernacular. The London critics had spoken well of David Hume's
works on philosophy, and Raeburn was painting his remarkable society
with unparalleled directness. The elite strove without much success to
acquire an English accent, but Adam Smith was envied because Balliol

273

College at Oxford had trimmed the roughness of his Fife dialect. The intelligentsia cultivated a thing called "rhetoric," which was supposed to be the canonical use of language freed from local vulgarities. And in the still-shabby Old College, a certain Mr. Hugh Blair was lecturing on the art of genteel conversation with much acceptance.

Into that world, a student named Joseph Black entered the university in 1750 for a medical degree. Born in France, he was a Scot in the true spirit of the word—honest, resolute, and obstinate, and he lived in a suburb of Auld Reekie. Twenty-seven years old and already a master of the chemical art, he had the advantage of a great teacher, William Cullen, who broached the subject of chemistry quantitatively as a branch of natural philosophy, with laws as fixed as those of mechanics. The most important lesson the student learned was the use of the chemical balance. Black graduated in 1754 after submitting as his dissertation his only major publication, a study of the acid humor that arose from *magnesia alba* ("white magnesia"; magnesium carbonate). No brief chemical essay has been so weighted with significant novelty. Everything about that remarkable work was historic!

FOR THE PAST SEVERAL YEARS, THE USE OF CAUSTIC REMEDIES FOR dissolving urinary bladder stones had come under scrutiny because of their adverse effects on body tissues. A new chemical called *magnesia alba*, which resembled mild alkalis such as chalk, had been introduced, and Black wanted to determine whether it might serve as a milder remedy for the stone. It did not! But his research gave him insights into the nature of air when he observed that a weighed amount of *magnesia alba* heated by fire *lost* weight. Moreover, like all weak alkalis, it gave off bubbles when treated with acid, indicating that it contained air. Black concluded that what was lost when heated must be an "air," and he called it "fixed air" (not an original term) because the common atmospheric air had somehow become "fixed" in the alkali. In fact, the term "fixed air" was used commonly for any sort of air contained in bodies.

Unknown to Black, his "fixed air" had already been described ear-lier by Van Helmont, who had named it *gas sylvestre*. Further experi-ments uncovered that fixed air was the main product given off during the fermentation of wine as well as when charcoal was burned, and it was present in the air exhaled by humans and animals. Combustion, fermentation, and respiration produced the same air. It was lethal to all creatures when breathed by itself. When driven through a clear solu-tion of limewater, it became milky—indeed, that soon became both a specific test for, and a property of, the fixed air (now known to be carbon dioxide, which forms a fine suspension of calcium carbonate in solu-tions of limewater, which is calcium hydroxide). It was among the first intensive and detailed studies of a chemical reaction.

Black was familiar with the effects of respiration upon atmospheric air. He knew about the reduction in its volume and the loss of its prop-erty of maintaining life when air was withdrawn using a vacuum pump. He became convinced that changes produced in inhaled atmospheric air during breathing involved a conversion of a portion of it into fixed air within the body. He recorded that since animals with more highly developed respiration were also warmer, it was probable that both an-imal heat and the formation of fixed air were somehow related and, perhaps, arose from the same process. The fixed air could not be atmo-spheric air but was another species that was present *within* the common air. Once again, attention was drawn to the fact that there could be many kinds of airs distinct from air.

Black went on to enunciate a theory of respiration that, his students confirmed, "though never published, is well known to us who have at-tended his lectures."[1] Breathing changed part of the common air to fixed air, which was accompanied by combustion that was generated in the lungs and then diffused through the body by Harvey's circulation. Fixed air was not a hypothetical entity like Boyle's "syderial or subterranean" substance or Hooke's "volatile salts" or Mayow's "nitro-aerial" particles. It possessed specific unique chemical properties that confirmed its pres-ence. Herein lay the great historical significance of Black's thesis: the *isolation* of a specific air and its characterization. The isolation and discovery of airs, and the discovery of the characters and laws of combination of gases, would be the focus of the coming century.

All such components were now called "airs" instead of "effervescences" or "vapors." Van Helmont coined the term *gas*, which would be popularized by Lavoisier.

SOUTH OF THE BORDER, LATE EIGHTEENTH-CENTURY ENGLAND WAS A paradise for the virtuoso, who was an amateur who loved his activities as well as remained independent of the Establishment. It was a time of discoveries, and the virtuosi were contributing. "Chemistry is the rage in London at present," John Playfair, the Scottish chemist and mathematician, recorded in his journal during a visit in 1782. Scientific experiment was suffused with an excitement and visionary breadth paralleled only by the contemporary Romantic poets. The virtuosi were exploring the constituents of air and water, the qualities of heat, and the new mysteries of electrical forces, all of which, perhaps, could build the better and freer society that the poets hoped for.

In witty England, the amateurism ran through everything: chemistry, philosophy, botany, and medicine. Much of the activity, also pervasive on the Continent, was superficial. England's George III dabbled in botany and Portugal's John V tried his hand at astronomy. Voltaire, who spent a vast amount of time weighing molten metal and cutting up worms, was also a dilettante. Wordsworth, Coleridge, and Shelley all owned microscopes at some point in their careers (Shelley had to pawn his when he eloped with Mary). Mary Shelley herself studied galvanic theory, which she applied to her novel creation: Frankenstein's monster. There was a freshness and freedom of mind and a new rash search for truth led by men and women who were rich enough, and grand enough, to do whatever they liked, who nonetheless did things that required a good deal of expertise.

Often, the eighteenth century's common man of science was bested by an intellectually superior woman. Voltaire's intimate companion, Madame du Châtelet, wrote on gravitation and translated Newton's

Principia; Diderot's great love, Mademoiselle Volland, was invariably a source of his own published ideas; and Madame de Pompadour pondered the motion of the stars. In a way, those amateurs were the heirs of the Renaissance ideal of the "universal" human. Some of them would build the foundation of modern science, and some of it would be profound.

KNOWLEDGE OF GASES OTHER THAN BLACK'S "FIXED AIR" HAD HITH-erto been lacking. Eight more years would pass before a stammering, eccentric nobleman and physicist Henry Cavendish would create nitric acid from the atmosphere using an electric spark. Cavendish, who could trace his descent back through six centuries to Sir John Cavendish in the reign of Edward III, was the most original and curmudgeonly of English scientists. He inherited a fortune in his middle age that made him one of the richest men in England but that had little influence upon his usual frugal habits. Overall, he cut a pathetic and ridiculous figure in society.

Having no knowledge on any subject except science and being too much out of touch with humanity—he was more blind to the human condition than unsympathetic to it—his social life revolved largely in science clubs and the meetings of the Royal Society. Other social inter-action was limited to weekly conversations with Royal Society members at the residence of its president, Sir Joseph Banks.

Cavendish's "singular oddities of character" and "extreme shyness" were well known. The sight of an unfamiliar face invariably sent him scurrying, and those who wished to speak to him were warned to avoid eye contact and to gaze into "the vacancy over his head." Any exchange with women was intolerable. He had a back staircase built at his home in Clapham to avoid chance encounters with scullery maids, and he or-dered his meals on slips of paper left on the hall table so that no speech with the housekeeper should be required. On his deathbed he was at-tended only by one valet, whom he sent away, asking not to be disturbed until a certain hour. When the valet returned, Cavendish was dead.

But his intellectual ability was of the highest order. He was one of the foremost chemists, mathematicians, and physicists of the day, with a masterly knowledge of the calculus. He would be best known as the scientist who determined the value of G, Newton's gravitational constant. He was the first to collect water-soluble gases over mercury, fully describe hydrogen, make a remarkably accurate calculation of the density of the earth, and discover the chemical composition of water (which he did, notably, before James Watt's more famous publication on the subject). It was Cavendish who first noticed the presence of water when common air and "inflammable air" (hydrogen) were sparked, which observation he related to his lifelong friend, the stammering Joseph Priestley, who then told Watt. Treatises found among his posthumous papers on heat and electricity as well as the mechanical and dynamic sciences heralded discoveries made a half century later.

Cavendish's success as a superb experimenter lay in his application of a new "balance" made exclusively for him by John Harrison, the lone carpenter and genius of chronometer fame who helped solve the problem of longitude at sea. With his new gadget, Cavendish was the first to determine the weights of equal volumes of gases. He next established an unequivocal connection between air and a component of the atmosphere for which the French chemist Chaptal had proposed the name "nitrogen." He then identified a second gas, insoluble in water and explosive when lit in the presence of common air, which he called "inflammable air," which was released when metals such as zinc, iron, or tin were dissolved in either "spirit of salt" (hydrochloric acid) or "oil of vitriol" (sulfuric acid). Another air, obtained from putrefying gravy, broth, and raw meat, he found to be a mixture of Black's "fixed air" and his "inflammable air," neither pure. In 1766, he sent a paper to the Royal Society titled "On Factitious Air," by which he implied gas produced artificially in the laboratory as distinct from the "natural" air.

ABOUT TWO MILLENNIA EARLIER, GREEK NATURAL PHILOSOPHERS had observed that fire, whether practical or mystical, caused changes in bodies, and when the flame was escaping, something was being lost and a relatively light ash left behind. Clearly, an inflammable principle escaped during combustion. More recently, in his own elemental theory, the Swiss alchemist Paracelsus had included fire as one of his three primary elements (*tria prima*). Then Jan Baptist van Helmont, a Flemish chemist and follower of Paracelsus and an acquaintance of Harvey, proposed the Greek word *phlogistos* (*phlox* means "flame") for the "matter of fire"—a particular "sulfurous earth" that was fixed within substances and that accounted for their combustibility. In 1703, Georg Ernst Stahl, a German chemist and physician, baptized the ancient inflammability principle with the catchy word "phlogiston," which he regarded as being "fixed" in substances. "This I was the first to call phlogiston!" he declared.[2] The theory had been proposed in the late 1600s by Stahl's teacher, a German experimenter, entrepreneur, and Paracelsian named Johann Joachim Becher.

Stahl had graduated at Jena and became physician to the Duke of Saxe-Weimar. Later, he was elected professor of medicine and chemistry at Halle and court physician to the King of Prussia. He was known to have a quarrelsome, bitter, and melancholic disposition; he rarely answered letters, showed contempt for all who differed from his views, and reacted violently to criticism. True to his nature, he interspersed his text, written in a combination of Latin and German, with alchemical symbols to intentionally make his works confusing and difficult to understand.

Phlogiston, Stahl settled in 1716, was the essential sulfurous component of substances that escaped during burning: "Fire [is] the momentous glance of phlogiston in its passage from one engagement to another," he wrote. It was present in large amounts in oils, fats, wood, charcoal, and other fuels. Combustible matter, Stahl held, consisted of that magic ingredient phlogiston plus a residue, called calyx, which was the ash. Sulfur, which burned leaving very little residue, was pure phlogiston.

The theory itself was seductively simple. Like all popular creeds and delusions, phlogiston had a plausible idea at its root. Metals, when heated in air, changed to calyx because they lost phlogiston. When the resulting calyx was heated with charcoal, it converted back into metal because it gained phlogiston from the charcoal, which was a fuel rich in phlogiston. Both observation and explanation were intuitively appealing. In his *Critique of Pure Reason*, Immanuel Kant lauded the phlogiston theory as a milestone in the progress of science. The mathematician Nicolas de Condorcet concluded, "If ever there was anything established in chemistry, it is surely the theory of phlogiston." It seemed to offer a unified explanation of combustion as well as the transformation of metals by heat (calcination) or by chemical combination. Phlogiston lingered for ninety years in the seventeenth and eighteenth centuries. In the end, it was as bad a theory as bad theories go, and as bad theories go, it went.

THE MEANING OF PHLOGISTON, AS USED BY STAHL, OFTEN APPROACHED the modern concept of "potential energy." Fire, in Stahl's theory, was not an element but an instrument that broke down matter into its component parts, a tool that decomposed mixtures by acting upon their phlogiston during burning, causing it to be released. When all phlogiston was consumed, the substance suffered a weight loss as it became "dephlogisticated." The common air, in turn, became "phlogisticated," that is, saturated with phlogiston to the point that it could no longer support combustion. The more phlogiston a substance contained, the greater would be its weight loss during combustion and the smaller would be the residue left behind. Substances that burned almost completely with little or no residue, like sulfur, were formed largely of phlogiston.

Stahl evidently thought that some clarification was necessary to explain why the atmosphere did not gradually become overwhelmed with phlogiston. Phlogiston released into the air, he reasoned, was reabsorbed by plants and trees, which is how wood and other inflammable natural substances acquired their own phlogiston. Thus, a natural cycle

for phlogiston existed in nature. Never mind that no one had seen, isolated, or analyzed phlogiston! It simply departed on combustion.

Here was a theory false to the verge of the ludicrous, a classic in the
history of scientific errors. Yet it appeared to integrate most facts familiar to chemists and even enabled the solution of new problems. Some
phlogistonists perceived a parallel between combustion and respiration,
just as some poets likened life to a flame. Respiration, like combustion,
involved a transference of phlogiston to the air, and both ceased in a
confined space when air became saturated with phlogiston.

Two airs were now differentiated: Black's "fixed air" and Cavendish's
"inflammable air." Could they, in some way, be related to the common
air? Were they modifications of it? Cavendish eventually regarded his
inflammable air (now known as hydrogen) to be a single, uniform substance that was a component of common air and that conformed to the
properties of Stahl's invention, phlogiston. Could inflammable air be
phlogiston? Cavendish expressed his findings characteristically: "I have
not indeed made sufficient experiments to speak quite positively on this
point." A year later, guided by his own experiments, Joseph Priestley
concluded that inflammable air was, indeed, phlogiston.

thirty-three

Concerto of Airs

*I soon realized that it was not possible to form
an opinion on the phenomenon of fire as long
as one did not understand air.*

—Carl Wilhelm Scheele

Joseph Priestley, the son of a wool-cloth maker and the discoverer of oxygen, was born of sturdy Puritan stock in the tiny village of Fieldhead at West Riding, in Yorkshire.[1] That was about all that he wished the world to know about himself. He would become a towering figure in British science after Harvey, and it would be impossible to ignore him. Honors were showered on two continents, including election to the Royal Society, which awarded him its Copley Medal (the Nobel Prize of that time), and he achieved the standing of a mythical hero after his death. His early years alone hinted that he was destined for extraordinary things when he was ordained for the ministry from the dissenting academy at Daventry, Northamptonshire. After ministering at Needham Market in Suffolk and at Nantwich in Cheshire, he eventually became a tutor at the prestigious Unitarian Warrington Academy in Lancashire.

His portrait shows a kindly face infused with neat, modest, middle-class sense. There is a certain quiet self-possession and benign composure visible; a gladness and reverence united in such harmony as bespeaks not a false or morbid state but a genuine healthy and robust mind. His father had willed him a sparse legacy, but not a penny of that inheritance ever reached its lawful owner, as the executors embezzled it. "The faith of faithless guardians," he wrote many years later, "plagued me from my boyhood." But loud or dramatic complaints were just not in young Joseph, who learned early how unjust and foolish his fellow mortals could be. A single man cannot amend the world's ways, he realized, and humanity's powers to prey upon and delude others and itself are limitless.

Like Demosthenes, he battled a serious speech impediment, and the more genteel members of his congregation objected to his stammering delivery. He was permanently afflicted by concern for money, to judge by his letters. Poverty invariably pressed hard upon him, and if he were to feel now and then that the fruits of his industry deserved more than he received in the way of material reward, he would be amply justified. He was, in addition, unpopular because of his Unitarian religious views. He seemed the amateur spokesman of all who labor at the edge of exhaustion, of the overburdened, of those who are already worn out but still hold themselves upright with scanty resources, who yet contrive by skillful husbanding and prodigious spasms of will to produce, at least for a while, the effect of gentility in poverty. There are many such unsung heroes and heroines in everyday life.

IN 1872, ASSERTING HIS STAND AGAINST RELIGIOUS DOGMA, PRIESTLEY published A History of the Corruptions of Christianity, an opinion not generally shared because it suggested that Jesus's claim to divinity had been foisted on him by a later (and lying) Saint Paul, just as all miracles had been invented by the Gospels. Thomas Jefferson, on the other hand, read Priestley's volume "over and over again; and I rest on them . . . as the basis of my own faith." Together with his other *An History of Early Opinions Concerning Jesus Christ*, he drew the hostility of Lutherans, Calvinists, and Anglicans alike, who all felt his works to be wicked, malignant, and destructive. Inevitably, with his meticulous assaults on the ideas of the Trinity, the Eucharist, predestination, the immateriality of the soul, and the Last Supper, and being a Unitarian in a nation of Anglicans under a Protestant monarch George III, he became an enemy of the state and a target of that mob violence and insurrection so splendidly described in Dickens's novel *Barnaby Rudge*.

Priestley's chapel, house, library, manuscripts, and apparatus were all burned by rioters and he himself was burned in effigy. In an open letter to "my late Townsmen and Neighbors," whom Priestley addressed after the carnage, he wrote: "You have destroyed my library . . . which no money can repurchase, except in course of time. But what I feel far more, you have destroyed manuscripts which have been the result of laborious study of many years, and which I shall never be able to recompense; and this has been done to one who never did, or imagined, you any harm." He closed, with little place for bitterness in his heart: "At all events we return you blessings for curses and hope that you shall soon return to that industry and those sober manners for which the inhabitants of Birmingham were formerly distinguished." Priestley could never get rid of that foolish hope that gentleness may yet win the assent of the brutish and the self-centered. Such blindness to the faults of men is at worst the weakness of an over-trustful nature.

In danger of his life, he took flight to London and then to America, to the serenity and wild seclusion of the banks of the Susquehanna, where the only sounds were the lowing of cattle and bleating of sheep and the ceaseless murmur of the stream. He knew that he was going to a far, far better place, where hundreds of the persecuted and impoverished had fled and where, he felt, science could be turned to better uses. He was not the first scientist-philosopher to seek safe harbor in the New World. Notwithstanding the blight that had befallen him, he built himself a home in the township of Northumberland in central Pennsylvania, 130 miles inland from Philadelphia, where other political refugees had settled. There he cobbled together a laboratory and a small library of leather-bound companions whom he read over and over, admitting a newcomer now and then after much deliberation.

For Joseph adored books. They offered company in solitude, not merely for the treasures hidden within that neither wealth could bribe nor artifice deceive but also for their material selves. He was a great, thoroughgoing bibliophile. Strange as it may seem, the mere touch of books gave him an enjoyable feeling of advancement in learning. Still keener was his passion for book hunting, and the two went well together. His moments of sheer happiness were those passed in the still ether of timeworn bookstores, and whenever he went on some far journey, he would turn aside to any shop that he chanced to see in the hope that some scrap of the writings he coveted might lie there. In his personal habits he was the most frugal of men, but he could not deny any income to be spent on the further acquisition of manuscripts. Just as the warrior is presented to us symbolically in helmet and armor and the bishop with his ring and crozier, so Priestley presents himself with the weapon of his own making: he is the man with the book.

HE COULD STILL MANAGE IN A PINCH, BUT AGE MADE THE PINCH harder. Though he had been, only a while earlier, a poor scholar who had eked out a subsistence by incessant toil, often giving private lessons to procure the wherewithal to put bread on the table, he was now wooed by the high and the learned. But the gentle feelings of those early days still abided. In the solitary backwoods of the New World, he never lost sight of the courtesies of polite life, and the gate to his home was forever swinging open to old and young alike. He remained the same substantial and determined yet meek and generally tolerant man. A lover of his timeworn, settled habits to the end, he stood contented by the ancient ways.

He was given a national welcome. The Tammany Society of New York sent a committee to receive him, the Unitarian Church offered him a ministry, and the University of Pennsylvania was ready to make him a professor. He accepted none. New feelings and new thoughts are understood only within a limited circle of the scholarly. On a visit to London in his earlier years, he had met "the old fox" Benjamin Franklin with whom he contracted a lifelong friendship; the two dined together nightly during the winter of 1774–1775 and now continued a voluminous correspondence. Thomas Jefferson consulted Priestley regarding the founding of the University of Virginia. He cemented a friendship with the physician Benjamin Rush as well as Jefferson's intellectual mentor, the astronomer David Rittenhouse. He left the wilderness on occasion to read before the American Philosophical Society, and he was a frequent guest at George Washington's table.

At his death of a hale and green old age on February 6, 1804, his home, with its books and burners, candles and crucibles, was dedicated by fellow chemists and admirers as a permanent memorial: the present Priestley House Museum. Nine years after his death, John Adams still acknowledged to Jefferson in their final letters that the "great, excellent, and extraordinary Man, whom I sincerely loved, esteemed, and respected, was really a Phenomenon: A Comet in the System" (Priestley was mentioned fifty-three times in their correspondence). Memorials were erected at Leeds, Northumberland, and Birmingham. In 1922, the American Chemical Society established the prestigious Priestley Medal. His complete writings, as edited by Joseph Rutt and published between 1817 and 1832, filled twenty-five volumes.

He would go down in the annals of science as a pioneer in the new field of pneumatic chemistry (the chemical analysis of gases), which evolved as a direct outcome of Harvey's demonstration of the circulation. Mayow had inferred from his experiments that "it clearly appears that animals exhaust the air of certain vital particles . . . that some constituent of the air absolutely necessary to life enters the blood in the act of breathing." What was that constituent? Here is where Priestley enters our story. He is the traditional discoverer of the "air" now called oxygen. Here was Galen's "vital spirit" reincarnated! Malpighi's capillaries had completed the anatomy of the circulation. Now Priestley's work set into motion the completion of the physiology of the circulation.

WHEN PRIESTLEY BEGAN AS LIBRARIAN TO LORD SHELBURNE AT Calne, his duties were slight, almost nominal, providing ample opportunity for experiments. He was, in common with the times, an amateur scientist. A self-taught chemist, he had acquired an extraordinary skill in manipulating gases that he applied to his chemical research, which would eventually be published in six volumes as *Experiments on Different Kinds of Air.*

Priestley noticed, among other things, that when he poured water from one tumbler into another within the layer of the "fixed air" above the vats of brewing beer, the water became impregnated with bubbles. The fizzy mixture even had a pleasant taste. Having failed to get interest from the Admiralty in his new water as a cure for scurvy, he lowered his sights. "Soda water," as it was called, rapidly became the rage of health spas throughout Europe and Priestley's name was made, but not his fortune (the English essayist G. K. Chesterton named it "windy water").

In 1772, Priestley, still a stammering nonconformist minister and a companion to Lord Shelburne, took a vital step forward by collecting all "airs" over mercury instead of water (a technique he learned from Cavendish). With that simple substitution, he isolated several new gases that were soluble in water, including nitrous oxide ("laughing gas"), hydrogen chloride, ammonia, sulfur dioxide, and nitrogen peroxide. The two tongue-tied geniuses, Cavendish and Priestley, individually lonely because of each's speech handicap, became kindred spirits and enjoyed a close friendship for over a quarter century. Cavendish published a new book, *Experiments on Air*, about his own experimentation with the common air that he often collected from his friend William Heberden's garden. Heberden would gain perennial fame in medicine as the discoverer of the potentially fatal clinical heart syndrome that he named "angina pectoris."

IN EARLY AUGUST 1774, PRIESTLEY ACQUIRED A LARGE BURNING LENS, twelve inches in diameter with a focal length of twenty inches, which had presumably belonged at one time to the Cosimo III de' Medici, Grand Duke of Tuscany. By heating with his new spyglass (as the lens was then called) the brick-red ash of mercuric oxide that was formed when mercury was heated in air, he isolated an "air" in which a candle burned with "dazzling splendor" and a piece of red-hot wood sparkled brilliantly before being rapidly consumed. Charcoal, too, instead of being expended quietly as it did in atmospheric air, burned with a flame attended with a crepitating noise and "threw out such a brilliant light that the eyes could hardly endure it." Mice lived longer in this new gas than in an equal volume of air. He regarded his new "air" as the purest part of the atmosphere and "highly respirable"; indeed, on taking a few whiffs of it himself, he fancied a certain sense of exhilaration, and suggested that it might be "peculiarly salutary" for the lungs in certain morbid conditions.

The following year, in a letter dated March 15, he reported his discovery to the Royal Society. It was the first public announcement of the isolation of a new air that could be safely breathed. Misled by the popular phlogiston theory, Priestley baptized his new gas with the awkward name "dephlogisticated air" because he reasoned that the superiority of the gas resulted from the loss of its phlogiston. He wrote:

My reader will not wonder that, after having ascertained the superior goodness of dephlogisticated air by mice living in it, and the other tests mentioned, I should have the curiosity to taste it by myself. I have gratified that curiosity by breathing it, drawing it through a glass siphon, and by this means I reduced a large jar full of it. The feeling of it to my lungs was not sensibly different from common air, but I fancied that my breast felt peculiarly light for some time afterward. Who can tell but that, in time, this pure air may become a fashionable article of luxury? Hitherto only two mice and I have had the privilege of breathing it.[2]

In that last sentence, Priestley (and his mice) convincingly claimed his priority in the discovery of the new gas, which was the first public announcement of the discovery of oxygen. There *was* an air purer than the common air. Shortly thereafter, he identified a second gas from the atmosphere that he simply called "phlogisticated air," which had all the properties of Black's "fixed air" (carbon dioxide).

AT THE SAME TIME, PRIESTLEY STUDIED THE EFFECT OF PLANTS ON the atmosphere as his older fellow clergyman and chemist Stephen Hales had done at his parsonage in the small country village of Teddington, situated fifteen miles from London between Hampton and Twickenham (where he had befriended the poet Alexander Pope). When Priestley placed a sprig of garden mint, and later balm, groundsel, and spinach, in a container whose air had been exhausted by a candle, he found that ten days later, not only was the mint thriving "in a most surprising manner" but a burning candle burned perfectly as well. He inferred: "Plants, instead of affecting the air in the same manner with animal respiration, reverse the effects of breathing, and tend to keep the atmosphere sweet and wholesome." It would be among his most important discoveries, better understood by posterity as the effect of sunlight on plants to produce oxygen through photosynthesis.

He shared his observations on the mint's capacity to rejuvenate "putrid" air with Benjamin Franklin, who was the first to grasp, however vaguely, the far-reaching process of food webs and ecosystems (a word coined by Oxford botanist Arthur Tansley in the 1930s) and the cycle of life on earth: plants converted light energy into chemical energy, absorbed carbon dioxide in the process, and gave out oxygen as a waste product, which, in turn, powered the animal world that released carbon dioxide as the waste—a natural cycle, which had originated about two billion years earlier in the Protozoic era.

In the paper "Observations on Respiration and the Use of Blood," published in 1776, Priestley suggested, for the first time, that the contact between air and blood in the lungs, which had been first described by Ibn al-Nafis and the fellow Unitarian Servetus, involved some *chemical* change, and thus advanced, a step further, the "purpose" of Harvey's circulation. Priestley proposed that the function of blood was not to receive phlogiston from the air, or to mix with phlogisticated air while it circulated, but rather to communicate phlogiston derived from the body's metabolism into the atmospheric air. When venous blood lost phlogiston during its passage through the lungs, its restoration to its florid state in the arteries was due to an addition of dephlogisticated air (oxygen) derived from the lungs: blood exposed to dephlogisticated air gave up its phlogiston and became bright red dephlogisticated blood. On the other hand, arterial blood exposed to phlogisticated air became dark and venous.

Although he was a radical in his religious and political beliefs, he was conservative in his scientific thinking and was never able to see his discovery through the contemporary mythology of phlogiston. Priestley neither impugned nor doubted the conclusions of phlogiston theory. Its reasoning seemed admirable, the deductions true, and the science deficient only in applicability. A new light was needed to bring about a clear and economical explanation outside the shutters of phlogiston, and it was precisely at that moment that a youthful Antoine Lavoisier came upon the scene.

thirty-four

Unto This Last

Longvil: But to what end do you weigh this Air, Sir?
Sir Nicholas Gimcrack: To what end shou'd I?
To know what it weighs! O knowledge is a fine thing.

— Thomas Shadwell

I F, ON THE PRACTICAL SIDE, THE STORY OF "AIRS" HAD BEGUN IN Scotland, the final drama was played out in France, specifically, in Paris. Antoine-Laurent Lavoisier suffered the advantage or handicap of being born into affluence.[1] The family wealth had become immense as he turned five, when his mother died. That good fortune undoubtedly played its part in establishing his place in the sun. His childhood, being educated largely at home, was secluded and solitary. He did not really "grow up" until he was almost twenty. The child with the fair hair who had clung to his late mother's affection was entrusted to an unremittingly zealous and sternly loving aunt for the right mixture of godliness and good learning, and a dotingly admiring and indulgent father, an advocate in the Paris *Parlement*, both of whom combined to make him a man of few intimates. Their guardianship was scarcely the best for him. He was allowed no wild adventures or rough exercises. They kept him forever by their side, wrapped up in blankets and well away from life's nasty accidents.

The pair were, one may say, like marsupials. They preserved him safely in a pocket, safe from rowdy companions and dangerous liaisons. The pocket was not altogether uncomfortable because it contained not only little Antoine but also the bulk of the family fortune as well. If the boy put his neck out a fraction of an inch too far, it was gently but firmly stuffed right back. Thus, his energies were confined within a high stockade of grim Christian righteousness. The guardian pair were far and away the predominant influence in his childhood: coddling yet stern, solicitous, exacting, indefatigable, and ubiquitous.

From his adolescence, he was pushed on every side to achievement, and achievement of no ordinary kind, for he could develop in the only directions that were permitted to him. And so, his young days never knew the sweet idleness that belongs to youth. He did feel a thrilling mixture of pleasure and apprehension when he received his independence, when he confronted his father with a defiant rejection of the whole Puritan principle that had emptied his childhood of games and toys.

Blessed in all his adult endeavors, his marriage, when he was twenty-eight to a girl of thirteen, Marie-Anne Paulze, proved to be the greatest asset to his career. It was an advantageous as well as an affectionate match. She took notes of her husband's experiments, edited and published his research, translated scientific articles written by the Oxford Chemists, sketched and engraved accurately composed illustrations of his experiments, and served as hostess in their luxurious home at fashionable 243 Boulevard de la Madeleine, which was open to French and foreign philosophers, writers, and scientists, including the two Josephs (Priestley and Black), and, above all, Benjamin Franklin, whom the Lavoisiers especially courted. Their portraits, painted by Jacques-Louis David, who was her tutor, show Marie-Anne as beautiful and serene, with a charm of the kind that emanates from lives fulfilled and fathomless and full of meaning. She, in turn, painted a portrait of her favorite Ben Franklin, which she presented as a gift in 1788 when he was the American ambassador at Paris. It is impossible to precisely define the quality of temperament and spirit that Franklin communicated to the Lavoisiers; "charm" affords the only explanation, and charm defies analysis.

LAVOISIER'S EARLY SCIENTIFIC PUBLICATIONS WERE LUCID, FORCE-ful, convincing, and, like nearly everything he wrote, intensely descrip-tive.[2] They excited pleasure and admiration, but not awe. Like Newton, Antoine was less the author of new experiments than the first to realize their full significance. Experimentation for its own sake, which had so delighted Priestley, had little appeal for him. He was a skillful analyst in the laboratory, which he equipped at his home with the most expen-sive and sophisticated technology in Europe (but intentionally made only in France), and he became an able exponent of the quantitative method like Black and Cavendish. But he was not greatly imaginative in the wild way that Cavendish was. Because of his precocious famil-iarity with science, he was elected to the elite Académie des Sciences at the age of twenty-five, where he enjoyed the company of much older men, including Franklin. The Académie was the ultimate arbiter of scientific progress in France.

At the birth of the Revolution, Lavoisier collaborated with the French National Assembly. But when violence and tyranny took over, he was stripped of the familiar assurances of aristocratic life and fell victim to the Reign of Terror. He was arrested on the accusation of one of his jealous former subordinates and convicted on a false charge on "24 fructidor of the Year One of the French Revolutionary Calendar." Another most virulent denunciator was Jean-Paul Marat, who hated La-voisier partly because, in his professional capacity at the Académie, he had contemptuously dismissed as worthless Marat's scientific work, a pamphlet titled *Research on Flame*. A doctor by profession, but entirely devoid of decent manners and with unwholesome associations, Marat was an undoubtedly ugly man suffering a skin disease that required frequent lying in a bathtub. Now Marat bitterly denounced Lavoisier as "this Coryphaeus of charlatans, sieur Lavoisier, son of a land grabber, chemical apprentice, pupil of a Genevese stock-jobber . . . the putative father of all discoveries which are noised abroad! Having no ideas of his own, he steals those of others."

Lavoisier was guillotined by his fatherland alongside his father-in-law in May 1794 at the age of fifty. The appeal judge was said to have remarked, "The Republic has no need of savants." When Lavoisier died, the Terror had only eleven more weeks to run. Marat would be famously assassinated in his bathtub by Charlotte Corday.

For some time, Lavoisier had vexed over the minuscule but definite *increase* in weight that occurred in the residual calyx when phosphorus or sulfur was burned in air, and tin and lead were calcined in a sealed vessel. He had reviewed thoroughly the available literature on phlogiston, including Priestley's work and the book by Priestley's colleague Richard Kirwan titled *An Essay on the Phlogiston*, all of which Marie-Anne translated and summarized.

It made little sense to Lavoisier that a substance should *gain* weight by *losing* phlogiston (or anything else), as was inherent in the phlogiston theory. It seemed more likely that combustible substances were absorbing something (either air or some component of air) to gain weight rather than releasing anything during combustion. In modern terms, oxygen was bonding to the heated material through oxidation by the flame of combustion. Moreover, though phlogiston was supposed to have weight, no one so far had managed to weigh that elusive (and nonexistent) entity.

Lavoisier had the right mindset and the right methods for isolating elements and then finding them again in compounds by weighing the proportions in which they combined. He deployed his chemical balance carefully and thoughtfully to determine the weights of substances that reacted together. "It can be taken as an axiom," he wrote, "that in every operation, an equal quantity of matter exists both before and after the operation." With that law of the indestructibility of matter (which had been described around 450 BCE by the Greek natural philosopher Anaxagoras), he set into motion the quantitative chemical laws and the direction of modern chemistry.

In 1772, at the age of twenty-nine, he recorded on a piece of paper, which he sealed:

About eight days ago I discovered that sulfur in burning, far from losing weight, on the contrary gains it; it is the same with phosphorus. . . . This discovery, which I have established by experiments, that I regard as decisive, has led me to think that what is observed in the combustion of sulfur and phosphorus may well take place in the calcination: and I am persuaded that the increase in weight of metallic calyxes is due to the same cause.[3]

That was the shot that would be heard throughout the scientific world. Lavoisier dismantled phlogiston theory and proposed a bold, new working hypothesis of combustion and respiration on a grand scale—without much evidence so far to support it! Something was taken up from the atmosphere during combustion. So, what was that something?

The more hard-core phlogiston enthusiasts defended their hypothesis by pointing out that, unlike all other masses, when phlogiston left a substance, it levitated upward, repelling Newton's force of gravity. Hence, phlogiston had negative weight! The logic went that the more phlogiston a substance contained, the lighter it became so that when phlogiston was lost, the substance *gained* weight! Others sought a way out by reducing the problem of mass to one of densities. Priestley, much wiser than most, simply said that, in physical science, mass was not always a consideration. He supported his thesis with three examples of newly discovered physical substances that had *not* been discussed in terms of mass or weight: the Newtonian ether, the electrical fluid of Benjamin Franklin, and the fluid of heat that Joseph Black had called the "caloric." Now Lavoisier explained the same facts in such a way that he swept away the encumbered phlogiston for all time.

IN THE FALL OF 1774, OCTOBER TO BE PRECISE, THE PAST AND THE future met and dined at Paris in the guise of two unlikely men who little realized where the years ahead would take them. One talked of the past, the other prepared for the future. The one looked ahead, the other looked back. One, roughly attired, was of a restless disposition and stuttered painfully as he explained himself to an audience of *philosophes*; the other was perfectly content to eagerly follow the halting monologue with a keen, ambitious face and a fired imagination. Just a few days earlier, one of them, the articulate and refined Lavoisier, had received a communication from an obscure Swedish apothecary named Carl Wilhelm Scheele, who claimed that, in the previous years, he had discovered that atmospheric air was a mixture of "several airs." One of its components, which he called *eldluft*, or "fire air," held the secret to combustion.

Scheele, the seventh of eleven children, was of German descent. He had been apprenticed to an apothecary at fourteen, during which time he taught himself chemistry from his master's books. He then moved to Malmo, Stockholm, and Uppsala, always in penury, before leasing a small apothecary shop from a widow in the smaller lakeside town of Koping. Here he lived a solitary life and, burdened by debt and depression, died in 1786 at the age of forty-three.

Sometime between 1771 and 1772, he noticed that by heating manganese oxide to the extreme, a gas was released, which he collected in an empty air bag and called "fire air." In it, powdered charcoal crackled brilliantly and burned rapidly. Scheele concluded that the ambient atmosphere was composed of two airs, which he called "foul air" and "fire air" (later identified as nitrogen and oxygen, respectively). From his book *Air and Fire*, published in obscurity in 1777, as well as from his posthumous papers, it was realized that he had, in addition, discovered chlorine and hydrogen sulfide; identified at least five inorganic acids and nine organic acids; isolated glycerin, arsenic, cyanide, and lactose; determined the nature of Prussian blue; and developed techniques for analyzing chemical fusion and separation! He would be recognized as one of the greatest of chemical experimenters and discoverers.

Now the stammering Priestley, for he was the other lecturing at the Paris dinner, informed Lavoisier of his own experiments with mercuric red oxide, performed just two months earlier as much by chance as by design. By burning the red calyx of mercury with his glass convex lens, he had produced an air that, too, supported combustion with extreme vigor. Here was the process in reverse! Instead of a metal being incinerated to form its oxide, Priestley had burned an oxide to extract the metal, just as in any smelting process. Little did Priestley realize that he had identified in 1774 and published that same year the same component that the obscure Swedish chemist Scheele had baptized as "fire air" in 1772 but published in 1777, which had also been described in 1774 by another French army apothecary named Pierre Bayen— altogether, the same "air" that Lavoisier, in 1779, would call le principe oxygine that we today call oxygen!

PRIESTLEY'S REVELATION USING RED MERCURIC OXIDE GAVE LAVOI-
sier an ingenious idea. Why not run the chemical experiment with this new red compound in both directions and measure exactly the quantities that were exchanged. First in the forward direction: burn mercury in a closed vessel and measure the ash and the exact amount of air taken up. Then run the process in reverse: take the same mercuric red oxide ash and heat it vigorously, as Priestley had done, to expel the air. Metallic mercury would be left behind and weighed, the released air could be collected in a vessel and measured, and the crucial question could then be answered: "How much?" Lavoisier performed the experiments with his highly precise equipment, even purchasing the red mercuric oxide ash from the same source that Priestley told him, and

the results were analyzed. They were identical to what Priestley had narrated. Only, Lavoisier examined the same facts without the veil of phlogiston preconception.

At the Easter meeting of the Académie des Sciences, Lavoisier disclosed that metals, when heated in air without charcoal, augmented their weight and gained a pure portion of the same air that surrounded man and that man breathed (that memorable document would eventually be called the Easter Memoir and be published with revisions in 1778). Lavoisier failed to mention that he had borrowed the idea of using red mercuric oxide as the source of oxygen from Priestley, making it implicit that he was the first and sole discoverer of oxygen! He had no intention of letting the credit go to anyone else, partly from self-glory and partly in the spirit of national chauvinism for French priority. When a Frenchman Edmond Genet challenged Lavoisier about the priority of Priestley's work, Lavoisier is said to have replied, "My friend, you know that those who start the hare do not always catch it."[4] Nonetheless, in a memoir that appeared in 1782, he did refer to oxygen as "the air which Mr. Priestley discovered about the same time as myself and even, I believe, before me," and subsequently, in his classic volume *Traité elémentaire de chimie* (*The elements of chemistry*), he remarked that it was discovered by Priestley, Scheele, and himself "almost simultaneously." He did address Priestley as "this celebrated physicist."

After abundant proof, Lavoisier named this portion the "oxygine" principle ("to generate acid"; Greek: *oxy* = acid, and *gen* = to beget). Wary of using a new word, he eventually preferred to call it "vital air." The truth of Lord Brougham's remark "Lavoisier never discovered oxygen until Priestley discovered it to him" can be accepted. But if clear grasp of its implication be accepted as the test of a discovery, then Lavoisier discovered oxygen. He was the first to define "oxygine" in terms of a theory embodied in facts.

LAVOISIER SPENT MOST OF HIS TIME BETWEEN 1772 AND 1778 DISMANtling the phlogiston theory. Immediately following his final experiments on calcination, he carried out analogous experiments showing that respiration was closely associated with the process of calcination. He had already noted that air entered into the "Composition of most Minerals, even Metals, and in very great abundance." Animating the air under a jar with his own breathing, he added to one part of the residue one-fifth part of "air" derived from the reduction of mercuric oxide. The resulting mixture appeared to him to be the common air. He could see at once that respired air resembled the air in which metals had been calcined insofar as it lost a certain quantity of the oxygine principle.

Air breathed out contained Black's "fixed air" (carbon dioxide) because the gas precipitated limewater, extinguished a flame, killed sparrows, and was unfit for breathing. There was no reason to doubt that Black's fixed air came from the lungs during breathing. The common atmospheric air must therefore be composed of at least two different substances: the oxygine, which was respirable and supported combustion and was capable of combining with metals on calcination to form their oxides, and the fixed air, which was nonrespirable and supported neither combustion nor calcination. To that Lavoisier added what he called "the noxious air which makes up three quarters of the atmospheric air and whose nature is still unknown to us" (now called nitrogen; Lavoisier called it *mofette*). Suddenly, air was no longer an element; it had become a compound, as Boyle and the Oxford Chemists had proposed in the previous century. Lavoisier inferred that the process of respiration involved two factors: atmospheric air was clearly modified in the lung so that "vital air" (oxygine) was being absorbed, and the lung substituted in its place an "equal" volume (later corrected to "less") of "fixed air."

Pursuing that line of investigation, he placed birds under bell jars to examine the effect of their respiration on the atmosphere.[5] Along the way, he was forced to notice a crucial difference between respiration and calcination. The air left by respiration precipitated limewater like Black's fixed air, whereas the air of calcination (in the absence of charcoal) caused no such change. Animals shut in a confined atmosphere succumbed as soon as they had converted the greater part of the vital oxygine into fixed air. The residual air left behind (nitrogen) was the same in calcination and respiration and was a purely passive portion that entered and left the lung unchanged. It could be reconverted into ordinary atmospheric air by simply adding oxygine. Augmenting or diminishing the quantity of oxygine in any atmosphere augmented or diminished the extent of time that an animal could survive in it.

From observations conducted and published in collaboration with the great mathematician Pierre-Simon Laplace, the Newton of France, Lavoisier concluded that respiration was a slow form of combustion like the combustion of carbon (charcoal), analogous to that operating in an oil lamp or a lighted candle. From that point of view, all creatures that breathed were combustible substances, burning and consuming themselves! Knowing that one property of oxygine was to communicate a red color—mercury with oxygine formed a red calyx—and observing an identical change in color of blood in the lungs, Lavoisier speculated that it was the oxygine that combined with blood in the lung and gave it its florid hue.

IT HAD TAKEN EIGHTEEN YEARS TO REACH THE MATURE THEORY OF 1790 from that day in 1772 when young Antoine had hastily scribbled a question on a folded sheet of loose paper: "But isn't the air composed of two substances, of which the lungs bring about the separation . . . of one of the two?" As important as Harvey's thought experiment on the "quantification" of blood was the quantification of air—the demonstration that an invisible, odorless, almost impalpable "something," bound up so closely with life and flame, was as strictly quantifiable as blood or water, or metals.

At the peak of his scientific career, Lavoisier summarized his physiologic studies in a final publication, *On the Respiration of Animals*, where he, in association with his young assistant chemist Armand Seguin, asserted that respiration was not limited only to the combustion of carbon but also combusted a part of the hydrogen in blood, which resulted in the formation of the gas carbonic acid and also water. In that final treatise on respiration, he voted in favor of the lung as the sole organ of respiration on the grounds of mere simplicity. Seguin participated as a "human guinea pig" in experiments in which his head was sealed within a mask and hood that connected to Lavoisier's apparatus via a long tube; a brass mask preserved in the Lavoisier family collection may well be the one actually used.

In 1789, Lavoisier published his *Traite elémentaire de chimie*, the world's first chemistry textbook. Its major contribution was the establishment, once and for all, of the concept of chemical "elements" in the modern sense, as substances that cannot be further decomposed—"simple radicals" is what they were called. Five years later, Lavoisier was guillotined. By this time, Scheele was already dead, in all probability poisoned by the fumes of his own furnace. The chemical "revolution" had coincided almost exactly with its political counterpart. Both had brought to birth forms of freedom for the world, and then straightaway had turned to compass the destruction of their own glorious offspring.

THE HOMERIC "WINE-DARK SEA" THAT HAD INSPIRED ALCMAEON'S ebb and flow of blood so long ago in 500 BCE on the Aegean shore reached its apotheosis in 1628 in Harvey's circulation in England and ended in Lavoisier's research on respiration in France at the end of the eighteenth century. Harvey had uniquely described the circulation, but even there, he could not break away from Aristotelian reasoning: that the circular movement allowed a preservation of the body—the microcosm—through a continual—circular—regeneration of blood. Circular motion served regeneration and preservation, and the circular motion of blood ensured cyclical regeneration of its virtue in the heart after its consumption in the organs. One hundred fifty years later, Lavoisier would be the first to explain the "purpose" of the Harveyan discovery and respiration and thermogenesis in animals.

What had driven those men, with such high courage and initiative as well as energy and patience, was not material rewards but a bold and determined purpose. Scholarship, research, and the spirit of science, perhaps one of the irreducible elements of the human spirit, on such a scale and thus conceived has in it something of the heroic! The most formidable of those movements was the discovery of the circulation. It was a paradigm shift, and historians have used it, with its corollary of early respiratory physiology, as a demarcation line to define the modern secular way of thinking about the microcosm in mechanical, biochemical, and molecular terms to reveal a hitherto unconceived complexity, exactness, and beauty in the human body.

What is now proved was once only imagined.

—WILLIAM BLAKE

CONCLUSION

*It is only when Death builds its frame around life
that the portrait of a man is really hung on a wall.*

—HENRY JAMES

thirty-five

In Our Time

Modern medical progress is the direct outcome
of the methods so successfully advocated and
practiced by William Harvey.

—H. B. Bayon

T HE PRESENT STORY, WHICH ENDS WITH LAVOISIER, SETS THE
stage for the development of our modern concepts and attitudes in
physiology and medicine, some of which are outlined here.[1] For over
a century, Harvey's novel idea had momentous theoretical significance
but lay unused, with hardly an immediate clinical application. That
difficulty had to await further advances in physics, chemistry, and technology before it could be applied to the complexities of the living body.

Giovanni Borelli, Malpighi's mentor, was among the first to apply
the better understanding of hydraulics offered by Galileo's physics and
mathematics to estimate theoretically the motive force of the heart as
a Harveyan pump.[2] Further research into blood pressure changes and
hemodynamics was initiated and described by the English preacher

Stephen Hales in the second volume of his 1733 treatise *Statical Essays, Containing Haemastaticks*.[3] Hales inserted a brass pipe into the "left crural artery" (femoral artery) of a mare "about 14 years of age . . . neither very lean nor yet lusty," and, by noting the height to which the blood rose in the pipe, measured the blood's pressure as "8 feet 3 inches perpendicular above the level of the left ventricle of the heart: but it did not attain to its full height at once." It was the first direct recording of arterial blood pressure in any species.

Clinical progress in understanding high blood pressure as a morbid condition began with the unveiling of its pathology in the kidneys and heart in autopsies of hypertensive patients with cloudy urine and "dropsy" (legs and abdominal swelling) in 1827 by the physician Richard Bright at Guy's Hospital, London (after whom Bright's disease is named). He observed that "the obvious structural changes in the heart have consisted chiefly of hypertrophy."[4] More clinical and autopsy observations in patients with different types of Bright's disease (large white kidney versus small white kidney, and so on) revealed hypertrophy of the muscle coats of the small arteries of the kidneys as well, and the presence in the kidneys of a "hyaline-fibroid formation in the walls of minute arteries, and a hyaline granular change in the corresponding capillaries." Those studies were conducted by other Guy's doctors, including Bright's contemporaries Thomas Addison and Thomas Hodgkin (of Addison's disease and Hodgkin's disease fame), whose combined investigations settled the basis of thinking about blood pressure as clinically "high" or "low." The possibility that the kidney changes primarily contributed to high blood pressure was raised.

The field was expanded by the brilliant Guy's physician Frederick Horatio Akbar Mahomed, who was the first to recognize and differentiate "primary hypertension" (now called "essential hypertension," which is the most common type of high blood pressure disease today) from secondary hypertension arising from kidney damage in Bright's disease. He observed that "previous to the commencement of any kidney change, or the appearance of any albumin in the urine, the first condition observable is high tension in the arterial system." Mahomed's vital observation, that high blood pressure is the cause of kidney (and heart) disease rather than vice versa, was validated by Huchard at Paris and Clifford Allbutt at Cambridge in the early twentieth century. Mahomed also introduced a practical sphygmograph for recording blood pressure from a disc placed on the radial artery at the wrist.

At about the same time, in 1846, the French physicist and physiologist Jean Leonard Marie Poiseuille began further experiments on blood pressure using fine glass tips inserted into arteries that were connected to a simple mercury manometer that he had invented. The following year, German physiologist Carl Ludwig added a practical float recorder that could trace variations in pressure on a rotating smoked drum, which allowed an appreciation of the physiological mechanisms that regulate blood pressure. The familiar inflatable arm cuff that we use today was introduced in 1896 by the Italian Scipione Riva-Rocci, who brought into general clinical use at the bedside an essential parameter of Harvey's circulation, that is, the force of contraction of the heart as a pump manifested as the blood pressure. That, in turn, allowed pioneering clinical research on hypertension to begin in 1914 by Franz Volhard and others and spawned modern blood pressure therapy in the mid-twentieth century. An accurate method of counting another Harveyan parameter, the arterial pulse rate, was introduced a couple of centuries earlier by Sir John Floyer, physician, author, and friend of Samuel Johnson.[5]

RICHARD LOWER'S GROSS UNRAVELING OF THE SPIRAL SETS OF MUSCLE in the heart, which confirmed Harvey's description of the heart as a pump, was recorded in his *De corde*. It was extended in the twentieth century by electron microscopy and scanning electron microscopy to study smaller and smaller heart muscle structures: from myofibrils to sarcomeres to molecules of the muscle proteins myosin and actin, and thence to peptide units and single atoms. Biochemically, it was recognized that calcium fluxes at the myosin-actin level contribute to heart pump function.

Harvey's brilliant quantitative thought experiment regarding the output of blood during each heartbeat awaited the new physics and mathematics under Galileo's (and later, Newton's) leadership to allow Borelli to calculate the force exerted by the limb muscles and to apply that finding, by analogy, to estimate the work of the heart as a piston (it turned out to be an erroneous estimate).[6] It would take another century before Ludwig's pupil Adolf Fick could calculate the cardiac output precisely by observing the difference between oxygen saturation in arterial and venous blood combined with the oxygen consumption per minute (called the Fick principle).

Cardiac catheterization, now a familiar diagnostic test that is possible only because the blood circulates, was first performed in 1844 by the eminent physiologist Claude Bernard to directly determine the temperatures of blood in the right and left ventricle, respectively, to refute Lavoisier's contention that animal heat was produced as a result of respiratory gas exchange in the lungs. If that were so, then blood entering the right ventricle and pulmonary artery should be cooler than blood in the pulmonary vein and left ventricle after its transit through the lungs. Bernard's later notebooks record measurements of intracardiac pressures in dogs and horses under a variety of physiological states and types of stimulation.

Further intracardiac pressure recordings were made in the 1860s by Auguste Chauveau and Étienne Marey in France to confirm Harvey's contention that it was the left ventricular apex during heart muscle contraction that was felt at the chest wall as the "apical impulse."[7] Marey especially noted that heart catheterization could be done safely. In 1898, German physiologist Nathan Zuntz validated the Fick method of measuring blood flow by introducing well-oiled (to prevent clotting) ureteral catheters into the ventricles of horses.[8] Subsequent technological advances extended its clinical application to humans in 1941 by the Nobel-winning pioneering work of André Cournand, which opened a new era, as cited by the Nobel Committee, in the clinical management of "respiratory and cardiocirculatory disease." The prize was shared with Cournand's mentor, the eminent Dickinson W. Richards, professor and chief of medicine at the Columbia division of Bellevue Hospital, and German urologist Werner Forssman, who had passed a ureteral catheter through a vein into his own right ventricle in 1929, confirming its position by X-ray and later by injecting a contrast dye, thus demonstrating the safety of heart catheterization in humans.

THE QUANTITATIVE ERA IN RESPIRATION PHYSIOLOGY IN TERMS OF OXygen and carbon dioxide gas exchange was launched by Lavoisier in the century after Harvey's death.[9] He demonstrated the chemical nature of respiratory gas exchanges in the lungs and showed them to be similar to the process of combustion, although he erroneously held the ultimate site of respiratory gas exchange with oxygen uptake, carbon dioxide elaboration, and heat production to be solely in the lung itself. A momentous advance was the simultaneous recognition by University of Pavia professor Lazzaro Spallanzani (who also published a critique of Homer's *Iliad*) that the absorption of oxygen and excretion of carbonic acid (carbon dioxide) was not confined to the lungs but occurred in different organs and tissues as well, such as brain, muscle, liver, and skin, thus inaugurating the era of "tissue respiration."[10] The research, conducted in the last decades of the eighteenth century during Lavoisier's lifetime, was published posthumously in 1803 as *Memoires sur la respiration*.

Supportive evidence was the forthcoming demonstration of higher oxygen and lower carbon dioxide contents in arterial than in venous blood. In 1861, German biochemist Moritz Traube concluded that "what we call respiration . . . represents the sum of those quantities of oxygen which are required by each organ . . . in this way, the respiration of, say, brain, liver, or spleen can be increased without the rest of the body organs showing a similar increase," a concept endorsed by his contemporary, physiologist E. F. M. Pfluger, and others, that respiration in all plants and animals was an intracellular biochemical process.

Noteworthy next steps were the discovery of hemoglobin as the precise site of utilization of oxygen transported by the arterial blood, the isolation of hemoglobin in the blood of the spleen, and investigation of the loose combination of oxygen with hemoglobin. The discovery of cytochromes by David Keilin and others in heart muscle cells is recorded in his superb book *The History of Cell Respiration and Cytochrome*, completed and published posthumously in 1966 by his daughter, Dr. Joan Keilin. The cytochromes marked the next major advance in cell respiration.

With the twentieth century came the metamorphosis of the heart as a pump into the heart as an engine that converts the energy of chemical reactions (metabolism) into mechanical energy, then into "work."[11] Having recognized the structural proteins of heart muscle, the next great question was how they might interact with energy metabolism. Biochemical mechanisms clearly lay at the heart of the problem, with both the energy-yielding processes and the motions of the engine occurring on the same scale of molecular dimensions in accordance with the laws of entropy.

The finding that tissue respiration can occur even in the absence of oxygen, demonstrated in 1824 by Jamaican-French physiologist William Fréderic Edwards and the German George Liebig, led to the appreciation that a vast number of cellular metabolic processes, including heart muscle contraction, are primarily a removal of hydrogen (dehydrogenation) rather than an addition of oxygen (oxidation). Hydrogen is transferred to a hydrogen acceptor, which can be oxygen but can also be something else. Moreover, dehydrogenation of metabolites such as lactate into pyruvate ends in the complete oxidation of such molecules to carbon dioxide and water, as demonstrated in the Nobel-winning research of Albert Szent-Györgyi and Hans Krebs, who discovered, respectively, that metabolites such as succinate, oxaloacetate, and others have a catalytic role in the main pathways of muscle contraction and are members of a cycle into which oxidizable matter enters and, in a single go-around, is re-formed, and the substrate molecule is converted to carbon dioxide and water (the Krebs cycle; well known to most high school science students).[12] The mechanism of Harvey's heart muscle contraction is now understood through molecular physiology.[13]

THE DUTCHMAN LEEUWENHOEK HAD NOTED THE VARIATIONS IN THE flow of red blood cells through capillaries, and that observation was confirmed and expanded qualitatively in the eighteenth century by the versatile Stephen Hales, the rector of Teddington and Alexander Pope's neighbor at Twickenham on the Thames. Hales, a polymath who had been the first to measure blood pressure in any species, demonstrated, a century after Harvey, that flow through a single capillary may vary from moment to moment as capillaries "are manifestly contracted or relaxed, by different Degrees of Warmth, Heat or Cold, or from the different qualities of the Fluids [chemicals] which pass through them."[14] Hales's experimental specimens of "capillary Arteries" most likely included arterioles, which have a muscular wall, as was made apparent by the finer microscopy of English physiologist Marshall Hall. Controversy on that subject raged throughout the eighteenth century because of the influential chemist Albrecht von Haller, who denied those capillary properties based on his own faulty and incomplete observations, until it was settled, independently, in the early twentieth century by the English investigators Thomas Lewis and Ernest Henry Starling and the Nobel-winning Dane, Auguste Krogh (assisted by his wife Marie Krogh).[15]

Investigations on the capillary blood pressure and the smaller vessels taken up by Poiseuille led to the finding that the "pressure and velocity of flow in the capillary vary . . . but [have] rather close relationship" (Poiseuille's equation). Independent quantification of diffusion, filtration, and absorption through capillary walls by Carl Ludwig, as well as by the Englishman Starling, confirmed that the real physiology of the grand Harveyan thesis lay in the physical laws that governed gas exchange across semipermeable membranes at the capillary level.[16] Their research, with later technological advances, would be applied clinically in dialysis machines and systems of extracorporeal circulation. In our own times, theory, experiment, speculation, and human ingenuity using isotope tracers and polarization microscopes continue to define how solvent and solute particles must pass through the capillary wall.[17]

The story of Harvey's circulation, and his unique description of the heart as a pump, promises to continue well into our new millennium beyond the striking advances cited above of our own lifetimes. For instance, adding an energy electron loss spectrometer (EELS) to scanning transmission electron microscopy and using aberration-corrected vibrational spectroscopy have the potential to allow cell metabolism in heart muscle cells to be examined in situ at the level of individual atoms.[18] And ballpoint pens with conductive "ink" from silver flakes dissolved in a polymer solution, which serve as semiconductors and dielectrics (a type of insulator), can now record and transmit the heart's electrical activity continuously through new "draw-on-skin" electronic circuits, without encumbering monitors.

Acknowledgments

THIS BOOK IS A WORK OF SYNTHESIS, NOT OF ORIGINAL SCHOLAR-ship. Though it is not written for specialized scholars but for the general reader of popular science and for healthcare providers who seek the proud heritage of their profession, it draws on the research of generations of experts, academics, historians, "natural philosophers," doctors, chemists, and painstaking commentators and reviewers who are represented here, invariably in their own words. Their publications are listed in the endnotes that form a mine of resources as well as a comprehensive bibliography. It is as much their book as mine, though some of the interpretations I have given to their work, or errors in the text, are my own. The 1847 translation of Harvey's works by Robert Willis, accessible in the public domain, is used in this narrative.

First of all, this book could not have been written without the long days and weekends of isolation endured by my wife, Avi. Writing a book nowadays is only part of the story. This work is in the hands of readers because of the excellent nurture of my agent, Jessica Papin at Dystel, Goderich & Bourret, who believed in a query from an unknown writer, steered me from query to signed contract in record time, and delivered me into the capable hands of my excellent editor Eric Henney at Basic Books. Jessica has been my guide, advisor, and companion through the whole process from query to book launch. It is Eric who came up with the book's title, which appeared to him in a dream! Great editors like Eric are what dreams of writers are made of! His superb guidance has unquestionably made the book what it is. Many thanks, too, to my editor Kyle Gipson who carried the manuscript forward, to Rebecca Lown for the cover design that captures the spirit of the book, to Christina Palaia for her diligent scrutiny, judicious deletions, and corrections, and to my production editor Kaitlin Carruthers-Busser, my publicist Jenny Lee, and all the unnamed good folks in their departments.

Additional Reading

A S A COROLLARY TO THE BIBLIOGRAPHY COMPILED IN THE END-notes, the following general histories offer excellent overall background to the specific history of the circulation, body heat, and respiration. They introduce readers to explore the history of medicine and cardiology. Useful resources are Garrison and Morton's *Medical Bibliography: An Annotated Checklist of Texts Illustrating the History of Medicine*, 2nd ed. (New York: Argosy Bookstore, 1954), and Logan Clendening's *Source Book of Medical History* (New York: Dover Publications, 1970). Invaluable are the two comprehensive volumes edited by W. F. Bynum and Roy Porter, *Companion Encyclopedia of the History of Medicine* (Routledge, 2013).

Ackerknecht, Erwin. *A Short History of Medicine*. Baltimore: Johns Hopkins University Press, 1982.

Bing, Richard J. *Cardiology: The Evolution of the Science and the Art*. New Brunswick, NJ: Rutgers University Press, 1999.

Bodemer, Charles, and Lester King. *Medical Investigation in Seventeenth Century England*. Los Angeles: William Andrews Clark Memorial Library, University of California, 1968.

Castiglioni, Arturo. A *History of Medicine*, translated by E. B. Krumbhaar. New York: Alfred A. Knopf, 1941.

Conant, James Bryant, ed. *Harvard Case Histories in Experimental Science*, Vols. 1–2. Cambridge, MA: Harvard University Press, 1964.

East, Terence. *The Story of Heart Disease*. London: William Dawson & Sons, 1958.

Garrison, F. H. *An Introduction to the History of Medicine*. Philadelphia: W. B. Saunders, 1929.

King, Lester. *The Medical World of the 18th Century*. Chicago: University of Chicago Press, 1958.

Lee, Thomas. *Eugene Braunwald and the Rise of Modern Medicine*. Cambridge, MA: Harvard University Press, 2013.

Major, Ralph. A *History of Medicine*, Vols. 1–2. Oxford: Blackwell Scientific, 1954.

Porter, Roy. *The Greatest Benefit to Mankind: A Medical History of Humanity*. New York: W. W. Norton, 2003.

Sigerist, Henry E. A *History of Medicine*, Vols. 1–2. Oxford: Oxford University Press, 1951.

Willius, Fredrick, and Thomas Dry. A *History of the Heart and the Circulation*. Philadelphia: W. B. Saunders, 1948.

Notes

INTRODUCTION

1. Thomas Kuhn, *The Structure of Scientific Revolutions* (Chicago: University of Chicago Press, 1962).

2. Ludwik Fleck, *Genesis and Development of a Scientific Fact* (Chicago: University of Chicago Press, 1979).

1. THE ORIENTAL HERITAGE

1. Charles Singer, *A Short History of Anatomy and Physiology from the Greeks to Harvey* (1925; repr., New York: Dover Publications, 1957), figs. 1 and 2.

2. Maoshing Ni, *The Yellow Emperor's Classic of Medicine: A New Translation of the Neijing Suwen with Commentary*, rev. ed. (Boston: Shambhala, 1995), 199; Huang Ti, *The Yellow Emperor's Classic of Internal Medicine*, trans. Ilza Veith (Berkeley: University of California Press, 1966).

3. H. R. Zimmer, *Hindu Medicine* (Baltimore: Johns Hopkins Press, 1948).

4. Zimmer, *Hindu Medicine*.

5. *Charaka Samhita*, 7 vols., trans. R. K. Sharma and Bhagwan Das (Varanasi, India: Chowkhamba Sanskrit Series Office, 2016).

6. Kaviraj Bhishgratna, *Susruta Samhita*, 3 vols. (Varanasi, India: Chaukamba Orientalia, 1907–1916; repr., 2004); G. D. Singhal and colleagues, *Susruta Samhita: Ancient Indian Surgery*, 3 vols. (Varanasi, India: Chowkhamba Sanskrit Series, 2007).

7. *Papyrus Ebers*, translated from the German version by Cyril P. Bryant (London: Geoffrey Bles, 1930; repr., Eastford, CT: Martino Fine Books, 2021), quoted in Ralph Major, *A History of Medicine*, Vol. 1 (Springfield, IL: Charles C. Thomas Publishers, 1954), 50; James Finlayson, "Ancient Egyptian Medicine," *British Medical Journal* 1 (1893): 748–752, 1014–1016, 1061–1064; B. Ebbel, *The Papyrus Ebers* (Copenhagen: Levin and Munksgaard, 1937); P. Ghalioungui, *The Ebers Papyrus* (Cairo: Academy of Scientific Research and Technology, 1987).

8. E. V. Boisaubin, "Cardiology in Ancient Egypt," *Texas Heart Institute Journal* 15 (1988): 80–85.

9. *Papyrus Ebers*, Bryant translation, quoted in Major, *A History of Medicine*, 1:46.

10. Henry James Breasted, Vol. 1 of *The Edwin Smith Surgical Papyrus*, translated in 2 vols. (Chicago: University of Chicago Oriental Institute Publications, University of Chicago Press, 1930; repr., Classics of Medicine Library, Birmingham, AL: Gryphon Editions, 1984), quoted in Fredrick Willius and Thomas Dry, *A History of the Heart and the Circulation* (New York: W. B. Saunders, 1948), 7; Major, *A History of Medicine*, 1:50; Arturo Castiglioni, *A History of Medicine* (New York: Alfred A. Knopf, 1941), 55–57.

11. M. A. Ruffer, *Studies in the Paleopathology of Egypt* (Chicago: University of Chicago Press, 1921); A. T. Sandison, "Degenerative Vascular Disease in the Egyptian Mummy," *Medical History* 6 (1962): 77–81; S. G. Shattock, "A Report upon the Pathological Condition of the Aorta of King Menephtah, Traditionally Regarded as the Pharaoh of the Exodus," *Proceedings of the Royal Society of Medicine* 2, no. 3 (Pathological sec.; 1909): 122–127; Tina Hesman Saey, "Mummies Reveal Hardened Arteries," *Science News*, September 6, 2014, 6–7.

2. THE SEA WITHIN US

1. Excellent resources on Greek medicine are James Longrigg, *Greek Rational Medicine* (London: Routledge, 1993); and James Longrigg, *Greek Medicine: From the Heroic to the Hellenistic Age* (London: Gerald Duckworth, 1998); unbeatable for Greek circulation physiology, to which this section of the narrative is heavily indebted, is C. R. S. Harris, *The Heart and the Vascular System in Ancient Greek Medicine: From Alcmaeon to Galen* (Oxford: Clarendon Press, 1973).

2. P. S. Codellas, "Alcmaeon of Croton, His Life, Work, and Fragments," *Proceedings of the Royal Society of Medicine* 25 (1932): 25–30.

3. Caroline Alexander, "A Winelike Sea," *Lapham's Quarterly*, Summer 2013, 201–208.

4. The term *physiologia* was first used by Jean Fernel, Vesalius's teacher in Paris. It is derived from the Greek *physis*, which loosely translated means "nature." It is used in this section in its modern sense.

5. Harris, *The Heart and the Vascular System*, 8.

3. SEEING IS BELIEVING

1. Ludwig Edelstein, "The Development of Greek Anatomy," *Bulletin of the History of Medicine* 3, no. 4 (1935): 235–248; H. von Staden, "The Discovery of the Body: Human Dissection and Its Cultural Contexts in Ancient Greece," *Yale Journal of Biology & Medicine* 65 (1992): 223.

2. Aristotle, *History of Animals*, trans. D'Arcy Wentworth Thompson (Oxford: Oxford University Press, 1910), repr., in *Britannica Great Books of the Western World*, Vol. 9 (Chicago: Encyclopaedia Britannica, 1952).

3. Aristotle, *History of Animals*. See also the discussion in Harris, *The Heart and the Vascular System*, 20–25; Singer, *A Short History of Anatomy and Physiology*, 10–11, fig. 10.

4. Aristotle, *History of Animals*, bk. 3, sec. 2–4.

5. Fritz Steckerl, *The Fragments of Praxagoras of Cos and His School*, collected, edited, and translated by Fritz Steckerl (Leiden: Brill, 1958), 11, 36, 588.

6. P. Potter, "Herophilus of Chalcedon: An Assessment of His Place in the History of Anatomy," *Bulletin of the History of Medicine* 50 (1976): 45; J. F. Dobson, "Herophilus of Alexandria," *Proceedings of the Royal Society of Medicine* 18 (1925): 19; L. L. Wiltse and T. G. Pait, "Herophilus of Alexandria (325–255 BC). The Father of Anatomy," *Spine* 23, no. 17 (1998): 1904–1914.

7. James Longrigg, "Anatomy in Alexandria in the Third Century BC," *British Journal for the History of Science* 21 (1988): 455–488.

8. G. A. Gibson, *Diseases of the Heart and Aorta* (New York: Macmillan, 1898), 174–175, quoted in Harris, *The Heart and the Vascular System*, 178–179.

4. THE LIFE IT BRINGS

1. Charles Singer, *A Short History of Anatomy and Physiology*, 10.

2. Daniel Graham, "A New Look at Anaximenes," *History of Philosophy Quarterly* 20 (2003): 1–20; John Burnet, *Greek Philosophy: Thales to Plato* (London: Macmillan, 1956).

3. J. R. Shaw, "A Note on the Anatomical and Philosophical Claims of Diogenes of Apollonia," *Apeiron* 11 (1977): 53.

4. Leonard G. Wilson, "Erasistratus, Galen, and the Pneuma," *Bulletin of the History of Medicine* 33 (1959): 293.

5. W. E. Leonard, *The Fragments of Empedocles* (Chicago: University of Chicago Press, 1905); see also E. R. Dodds, *The Greeks and the Irrational* (Berkeley: University of California Press, 1951), 146.

6. Diogenes Laertius, *Lives of Eminent Philosophers*, trans. R. D. Hicks, Loeb Classical Library (Cambridge, MA: Harvard University Press, 1925), 8:62. Also quoted in James Longrigg, *Greek Medicine*, 35.

7. Matthew Arnold, "Empedocles on Etna," in *Empedocles on Etna, and Other Poems* (n.p., 1852).

8. N. B. Booth, "Empedocles' Account of Breathing," *Journal of Hellenic Studies* 80 (1960): 10.

9. Aristotle, *On Respiration*, trans. W. Ogle (London: Longmans, Green, and Company, 1897), 7, 474, 17–20; W. Ogle, *Aristotle on Youth & Old Age, Life & Death, and Respiration*, translated, with introduction and notes (London: Longmans, Green, and Co., 1897); quoted in Harris, *The Heart and the Vascular System*, 16.

10. T. D. Worthen, "Pneumatic Action in the Klepsydra and Empedocles' Account of Breathing," *Isis* 61 (1970): 520.

11. Heather Webb, *The Medieval Heart* (New Haven, CT: Yale University Press, 2010), 92.

12. F. M. Cornford, *Plato's Cosmology: The "Timaeus" of Plato Translated with a Running Commentary* (New York: Routledge, 1935), 79a–e; Harris, *The Heart and the Vascular System*, 119–120.

13. Plato, *Timaeus*, 79a.

14. Aristotle, *On Breath*, trans. W. S. Hett (London: William Heinemann, 1957), 88, 474a15, 25–28. Also printed in Loeb Classical Library (Cambridge, MA: Harvard University Press, 1957).

15. Heather Webb, *The Medieval Heart*, 96.

5. BODY FLAME

1. Excellent sources for the evolution of the theory of animal heat are Everett Mendelsohn's *Heat and Life* (Cambridge, MA: Harvard University Press, 1964); and G. J. Goodfield, *The Growth of Scientific Physiology, Physiological*

Method and the Mechanist-Vitalist Controversy, Illustrated by the Problems of Respiration and Animal Heat (London: Hutchison, 1960).

2. Frank R. Hurlbutt Jr., "Peri Kardies: A Treatise on the Heart from the Hippocratic Corpus: Introduction and Translation," *Bulletin of the History of Medicine* 7 (1939): 1111, 1113.

3. Heather Webb, *The Medieval Heart*, 102.

4. Jerome Bylebyl, ed., *William Harvey and His Age: The Professional and Social Context of the Discovery of the Circulation* (Baltimore: Johns Hopkins University Press, 1979), 52.

5. Friedrich Solmsen, "The Vital Heat, the Inborn Pneuma and the Aether," *Journal of Hellenic Studies* 77, pt. 1 (1957): 119–123.

6. Thomas S. Hall, *History of General Physiology*, Vol. 1 (Chicago: University of Chicago Press, 1975), 33.

7. Aristotle, *De generatione animalium*, trans. Arthur Platt (1902), repr., *Britannica Great Books of the Western World*, Vol. 9 (Chicago: Encyclopaedia Britannica, 1952); also translated by A. L. Peck, in Loeb Classical Library (Cambridge, MA: Harvard University Press, 1943).

8. Fritz Steckerl, *The Fragments of Praxagoras of Cos and His School*, 11, 36, 588.

6. THROUGH THE BLOOD DARKLY

1. Plato, *Timaeus*, 70a–d.

2. Hippocrates, *On the Localities in Man*, trans. Francis Adams (1886); repr., *The Genuine Works of Hippocrates*, trans. Francis Adams, Books That Changed the World series (New York: Easton Press, 2005); quoted in Alfred Fishman and Dickinson W. Richards, eds., *Circulation of the Blood: Men and Ideas* (New York: Springer-Verlag, 1982), 6.

3. Heinrich von Staden, *Herophilus: The Art of Medicine in Early Alexandria* (Cambridge: Cambridge University Press, 1989).

7. KNOT OF VEINS

1. Frank R. Hurlbutt Jr., "Peri Kardies: A Treatise on the Heart from the Hippocratic Corpus," *Bulletin of the History of Medicine* 7 (1939): 1104, 1939; J. Wiberg, "The Medical Science of Ancient Greece: The Doctrine of the Heart," *Janus* 41 (1937): 225.

2. James Rochester Shaw, "Models for Cardiac Structure and Function in Aristotle," *Journal of the History of Biology* 5 (1972): 355; Arthur Platt, "Aristotle on the Heart," in *Studies in the History and Method of Science*,

Vol. 2, ed. Charles Singer (Oxford: Clarendon Press, 1921), 521; Harris, *The Heart and the Vascular System*, 123–134.

3. R. Van Praagh and S. Van Praagh, "Aristotle's 'Triventricular' Heart and the Relevant Early History of the Cardiovascular System," *Chest* 84 (1983): 462; T. H. Huxley, "On Certain Errors Respecting the Structure of the Heart Attributed to Aristotle," *Nature* 21 (1880): 1–5.

4. R. V. Christie, "Galen on Erasistratus," *Perspectives in Biology and Medicine* 30 (1987): 440; Kenneth D. Keele, "Three Early Masters of Experimental Medicine—Erasistratus, Galen and Leonardo da Vinci," *Proceedings of the Royal Society of Medicine* 54 (1961): 577–588; John Scarborough, "Erasistratus: Student of Theophrastus?" *Bulletin of the History of Medicine* 59 (1985): 515.

5. James Longrigg, *Greek Medicine*, 62.

6. Harris, *The Heart and the Vascular System*, 19–21.

7. An excellent scholarly account of the heart-brain primacy controversy is detailed by Heather Webb in *The Medieval Heart*.

8. Quoted in Jerome Bylebyl, *William Harvey and His Age*, 53–54.

9. Bylebyl, *William Harvey and His Age*, 53–54.

9. PRINCE OF PHYSICIANS

1. The best modern biography of Galen is Susan P. Mattern, *The Prince of Medicine: Galen in the Roman Empire* (Oxford: Oxford University Press, 2013); see also J. S. Prendergast, "The Background of Galen's Life and Activities, and Its Influence on His Achievements," *Proceedings of the Royal Society of Medicine* 23 (1930): 1131–1148.

2. Vivian Nutton, "Galen in the Eyes of His Contemporaries," *Bulletin of the History of Medicine* 58 (1984): 315.

3. R. J. Hankinson, ed., *The Cambridge Companion to Galen* (Cambridge: Cambridge University Press, 2008).

4. Rudolph E. Siegel, *Galen's System of Physiology and Medicine* (Basel: Springer Karger, 1968); Owsei Temkin, "On Galen's Pneumatology," *Gesnerus* 8 (1951): 180.

10. COMING TO GRIPS

1. Rudolph E. Siegel, *Galen's System of Physiology and Medicine* (Basel: Springer Karger, 1968). Important discussions of the heart, lungs, and respiration were accumulated by Galen in his *On Anatomical Procedures* and

On the Usefulness of the Parts (*De usu partium*), which he called a sacred discourse composed "as a true hymn of praise to our Creator." Four treatises were devoted specifically to aspects of the blood and respiration: *On the Value of Respiration, On the Causes of Respiration, On the Function of the Pulse*, and *Whether Blood Is Contained in the Arteries in Nature*.

2. J. C. Davies, "Galen and Arteries," *British Medical Journal* 28 (1970): 567.

3. Rudolph E. Siegel, "Galen's Experiments and Observations of Pulmonary Blood Flow and Respiration," *American Journal of Cardiology* 10 (1962): 738; Donald Fleming, "Galen on the Motions of the Blood in the Heart and Lungs," *Isis* 46 (1955): 14; Seigel, *Galen's System of Physiology and Medicine*.

4. Siegel, *Galen's System of Physiology and Medicine*.

5. Galen, *On the Natural Faculties*, bk. 3, trans. A. J. Brock (Cambridge, MA: Harvard University Press, 1915), 303–323; repr., *Britannica Great Books of the Western World*, Vol. 7 (Chicago: Encyclopaedia Britannica, 1952).

6. Owsei Temkin, "On Galen's Pneumatology," *Gesnerus* 8 (1950): 180–189.

7. Donald Fleming, "Galen on the Motions of the Blood in the Heart and Lungs," 14; the argument is presented here that Galen used the metaphor of ebb and flow to illustrate the kind of absurdity that nature would *not* fall into; see also A. R. Hall, "Studies in the History of the Cardiovascular System," *Bulletin of the History of Medicine* 34 (1960): 301–413.

11. BRAVE NEW WORLD

1. Virtually any book by Martin Kemp is an outstanding biographical source for Leonardo.

2. Charles D. O'Malley and John B. deCusance Morant Saunders, trans., *Leonardo da Vinci on the Human Body* (New York: Gramercy Books, 1982); Kenneth D. Keele, *Leonardo da Vinci on Movement of the Heart and Blood* (London: Harvey and Blythe, 1952); E. Belt, *Leonardo the Anatomist* (Lawrence: University of Kansas Press, 1956).

3. O'Malley and Saunders, *Leonardo da Vinci on the Human Body*, 286.

4. O'Malley and Saunders, *Leonardo da Vinci on the Human Body*, 224.

5. Edward MacCurdy, ed., *The Notebooks of Leonardo da Vinci* (New York: Reynal and Hitchcock, 1939), 128.

12. THE PROPER STUDY OF MAN

1. The best biographies in English are Stephen N. Joffe, *Andreas Vesalius: The Making, the Madman, and the Myth* (AuthorHouse, 2014); J. B. de C. M. Saunders and Charles O'Malley, *Andreas Vesalius* (New York: Butterworth-Heinemann, 1950).

2. Joffe, *Andreas Vesalius*, 56.

3. Samuel W. Lambert, Willy Wiegand, and William M. Ivins Jr., *Three Vesalian Essays to Accompany the Icones Anatomicae of 1934* (New York: Macmillan, 1952).

4. Charles Singer, *New Worlds and Old* (London: William Heinemann Medical Books, 1951), 7.

5. W. F. Bynum and Roy Porter, *Companion Encyclopedia of the History of Medicine* (London: Routledge, 1993), 86–87.

6. Michael Foster, *Lectures on the History of Physiology During the Sixteenth, Seventeenth and Eighteenth Centuries* (Cambridge: Cambridge University Press, 1901), lecture 1:14; John Farquhar Fulton, *Selected Readings in the History of Physiology* (Springfield, IL: Charles C. Thomas, 1930).

7. W. F. Bynum and Roy Porter, *Companion Encyclopedia of the History of Medicine*, 86–87.

13. A DISINHERITED MIND

1. Stephen N. Joffe, *Andreas Vesalius*, 131.

2. Charles Singer, "Some Galenic and Animal Sources of Vesalius," *Journal of the History of Medicine and Allied Sciences* 1 (1946): 6–24.

3. Michael Foster, *Lectures on the History of Physiology*, lecture 1:18.

14. THE MEDICAL MERCHANTS OF VENICE

1. The scholarly source for pre-Vesalian anatomy, which is the subject of this chapter, is L. R. Lind, *Studies in Pre-Vesalian Anatomy: Biography, Translations, Documents* (Philadelphia: American Philosophical Society, 1975).

2. Alessandro Benedetti, *Anatomice*, bk. 3, chap. 8, in Lind, *Studies in Pre-Vesalian Anatomy*, 106.

3. Niccolo Massa, *Introductory Book on Anatomy*, chaps. 17, 28, in Lind, *Studies in Pre-Vesalian Anatomy*, 214–218.

4. Erwin Ackerknecht, "Primitive Autopsies and the History of Anatomy," *Bulletin of the History of Medicine* 13 (1943): 334–339. A good discussion of Mondino's *Anathomia* is by Heather Webb, *The Medieval Heart*, 148–153.

5. L. R. Lind, *A Short Introduction to Anatomy (Isagogae Breves)*, translated and with an introduction and historical notes (Chicago: University of Chicago Press, 1959); Sanford V. Larkey and Linda Tum Suden, "Jackson's English Translation of Berengarius da Carpi's *Isagogae Breves*, 1660 and 1664," *Isis* 21 (1934): 57–70.

6. E. W. Le Gros Clarke, "Berengario da Carpi," *St. Thomas's Hospital Gazette* (London), 32 (1929–1930): 110–121; Lynne Thorndike, "Anatomy from Carpi to Vesalius," in *A History of Magic and Experimental Science*, Vol. 5 (New York: Columbia University Press, 1941), chap. 13; K. F. Russell, "Jacopo Berengario da Carpi," *Australian and New Zealand Journal of Surgery* 23 (1953): 70–72.

7. Lind, *A Short Introduction to Anatomy*, 96.

15. HUNTED HERETIC

1. Modern biographies of Servetus are Roland H. Bainton, *Hunted Heretic* (Boston: Beacon Press, 1953); L. Goldstone and N. Goldstone, *Out of the Flames* (New York: Broadway Books, 2003); see also William Osler, "Michael Servetus," *Bulletin of the Johns Hopkins Hospital* 21 (1910): 1–11.

2. M. Servetus, *The Restoration of Christianity*, trans. C. A. Hoffman and M. Hillar (Lewiston, NY: Edwin Mellen Press, 2007).

3. Michael Foster, *Lectures on the History of Physiology*, lecture 1:23; and John Farquhar Fulton, *Selected Readings in the History of Physiology*, 43; the extract on the pulmonary circulation is reproduced in Charles D. O'Malley, *Michael Servetus: A Translation of His Geographical, Medical and Astrological Writings with Introductions and Notes* (Philadelphia: American Philosophical Society, 1953), 201–208; and in Mark Graubard, *Circulation and Respiration: The Evolution of an Idea* (New York: Harcourt, Brace, and World, 1964), 83; as well as in Alfred Fishman and Dickinson W. Richards, *Circulation of the Blood*, 21–22.

4. Foster, *Lectures on the History of Physiology*, lecture 1:23; Fulton, *Selected Readings in the History of Physiology*, 43; O'Malley, *Michael Servetus*, 201–208; Graubard, *Circulation and Respiration*, 83; Fishman and Richards, *Circulation of the Blood*, 21–22.

5. Foster, *Lectures on the History of Physiology*, lecture 1:23; Fulton, *Selected Readings in the History of Physiology*, 43; O'Malley, *Michael Servetus*,

201–208; Graubard, *Circulation and Respiration*, 83; Fishman and Richards, *Circulation of the Blood*, 21–22.

6. Stephen Mason, *A History of the Sciences* (New York: Collier Books, 1962), 219.

16. A EUREKA MOMENT

1. R. S. Tubbs, S. Linganna, and M. Loukas, "Matteo Realdo Colombo (c. 1516–1559): The Anatomist and Surgeon," *American Surgeon* 74 (2008): 84–86; J. W. Hurst and W. B. Fye, "Realdo Colombo," *Clinical Cardiology* 25 (2002): 135–137.

2. Michael Foster, *Lectures on the History of Physiology*, lecture 2:28–30; R. J. Moes and C. D. O'Malley, "Realdo Colombo: 'On Those Things Rarely Found in Anatomy': An Annotated Translation from the *De Re Anatomica* (1559)," *Bulletin of the History of Medicine* 34 (1960): 508–528; more extracts are reproduced in Alfred Fishman and Dickinson W. Richards, *Circulation of the Blood*, 24; Mark Graubard, *Circulation and Respiration*, 93.

3. Foster, *Lectures on the History of Physiology*, lecture 2:28–30; Moes and O'Malley, "Realdo Colombo," 508–528; Fishman and Richards, *Circulation of the Blood*, 24; Graubard, *Circulation and Respiration*, 93.

17. ARABIAN KNIGHT

1. C. D. O'Malley, "A Latin Translation of Ibn Nafis (1547) Related to the Problem of the Circulation of the Blood," *Journal of the History of Medicine and Allied Sciences* 12 (1957): 248–253.

2. P. Ghalioungui, "Was Ibn al-Nafis Unknown to the Scholars of the European Renaissance?" *Clio Medicine* 18 (1983): 37.

3. R. E. Abdel-Halim, "Contributions of Ibn Nafis (1210–1288 AD) to the Progress of Medicine and Urology," *Saudi Medical Journal* 29 (2008): 13–22.

4. S. I. Haddad and A. A. Khairallah, "A Forgotten Chapter in the History of the Circulation of the Blood," *Annals of Surgery* 104 (1936): 1–8. Commentary and quotations by Max Meyerhof, "Ibn an-Nafis (XIIIth cent.) and His Theory of the Lesser Circulation," *Isis* 23 (1935): 100–120, reproduced in Mark Graubard, *Circulation and Respiration*, 59.

5. Haddad and Khairallah, "A Forgotten Chapter," 1–8; Meyerhof, "Ibn an-Nafis," 100–120, in Graubard, *Circulation and Respiration*, 59.

6. E. G. Browne, *Arabian Medicine* (Cambridge: Cambridge University Press, 1921).

7. J. B. West, "Ibn al-Nafis, the Pulmonary Circulation, and the Islamic Golden Age," *Journal of Applied Physiology* 105 (2008): 1877–1880.

18. WHO'S ON FIRST?

1. L. G. Wilson, "The Problem of the Discovery of the Pulmonary Circulation," *Journal of the History of Medicine and Allied Sciences* 17 (1962): 229–244; Owsei Temkin, "Notes and Comments: Was Servetus Influenced by Ibn an-Nafis?" *Bulletin of the History of Medicine* 8 (1940): 731–734.

2. Leonard L. Mackall, "A Manuscript of the *Christianismi Restitutio* of Servetus, Placing the Discovery of the Pulmonary Circulation Anterior to 1546," *Proceedings of the Royal Society of Medicine* (Section of the History of Medicine) 17 (1923): 35–38.

3. Quoted in Alfred Fishman and Dickinson W. Richards, *Circulation of the Blood*, 25.

4. J. P. Arcieri, *The Circulation of the Blood and Andrea Cesalpino of Arezzo* (New York: S. F. Vanni, 1945).

5. W. B. Fye, "Andrea Cesalpino," *Clinical Cardiology* 19 (1996): 969–970.

6. Quoted in Mark Graubard, *Circulation and Respiration*, 103.

19. WILD SEA OF TROUBLES

1. Gweneth Whitteridge, *William Harvey and the Circulation of the Blood* (New York: Neale Watson Academic Publications, 1971), 58–68.

2. Jerome Bylebyl and Walter Pagel, "The Chequered Career of Galen's Doctrine on the Pulmonary Veins," *Medical History* 15 (1971): 211.

3. J. P. Arcieri, *The Circulation of the Blood and Andrea Cesalpino*, 427–428.

20. PRELUDE TO GLORY

1. Excellent biographies include Louis Chauvois, *William Harvey: His Life and Times; His Discoveries; His Methods* (New York: Philosophical Library, 1957); Geoffrey Keynes, *The Life of William Harvey* (Oxford: Clarendon Press, 1978); Kenneth D. Keele, *William Harvey: The Man, the Physician, and the Scientist* (New York: Thomas Nelson and Sons, 1965); D'Arcy Power, *William Harvey* (1897; repr., New York: Heirs of Hippocrates Library, 1995).

2. E. A. Underwood, "The Early Teaching of Anatomy at Padua," *Annals of Science* 19 (1963): 1–26.

3. P. Palmieri, "Science and Authority in Giacomo Zabarella," *History of Science* 45 (2007): 404–442; H. Mikkeli, *An Aristotelian Response to Renaissance Humanism: Jacopo Zabarella on the Nature of Arts and Sciences* (Helsinki: Finnish Historical Society, 1992).

4. An unequaled study of Harvey's philosophical background and his heavy dependence on Aristotle is R. K. French, *William Harvey's Natural Philosophy* (Cambridge: Cambridge University Press, 1994); see also Walter Pagel, *William Harvey's Biological Ideas: Selected Aspects and Historical Background* (Basel: S. Karger, 1967); and Chauvois, *William Harvey: His Life and Times*.

5. J. Prendergast, "Galen's View of the Vascular System in Relation to That of Harvey," *Proceedings of the Royal Society of Medicine* 21 (1921): 1839–1848; F. G. Kilgour, "Harvey's Use of Galen's Findings in His Discovery of the Circulation of Blood," *Journal of the History of Medicine and Allied Sciences* 12 (1957): 232–234.

6. F. D. Zenman, "The Old Age of William Harvey," *Archives of Internal Medicine* 3 (1963): 829.

7. Quoted in Chauvois, *William Harvey: His Life and Times*, 168.

8. A. Pazzini, "William Harvey, Disciple of Girolamo Fabrizi d'Aquapendente and the Paduan School," *Journal of the History of Medicine* 12 (1957): 197–201.

9. A. H. Scultetus, J. Leonel Villavicencio, and Norman M. Rich, "Facts and Fiction Surrounding the Discovery of the Venous Valves," *Journal of Vascular Surgery* 33 (2001): 435–441.

10. J. P. Arcieri, *The Circulation of the Blood and Andrea Cesalpino*, 77–79, 81, 117.

11. Quoted in Michael Foster, *Lectures on the History of Physiology*, lecture 2:36–37; Hieronymus Fabricius, *Valves of Veins*, trans. K. J. Franklin (Springfield, IL: Charles C. Thomas, 1933), 47–56.

21. THE FINEST HOUR

1. Scholarly reviews of Harvey's methods in deducing the circulation include Gweneth Whitteridge, *William Harvey and the Circulation of the Blood* (New York: Neale Watson Academic Publications, 1971), which is the classic statement of the view that Harvey was a modern thinker. See also R. K. French, *William Harvey's Natural Philosophy* (Cambridge: Cambridge University Press, 1994); Walter Pagel, *William Harvey's Biological Ideas: Selected Aspects and Historical Background* (Basel: S. Karger, 1967); and Louis

Chauvois, *William Harvey: His Life and Times; His Discoveries; His Methods* (New York: Philosophical Library, 1957).

2. I. Bernard Cohen, *Revolution in Science* (Cambridge, MA: Belknap Press of Harvard University Press, 1985), 187–194.

3. Thomas Fuchs, *Mechanization of the Heart: Harvey and Descartes*, trans. Marjorie Grene (Rochester, NY: University of Rochester Press, 2001), 46.

4. Don G. Bates, "Harvey's Account of His 'Discovery,'" *Medical History* 36 (1992): 361–376; Lord Cohen of Birkenhead, "The Germ of an Idea, or What Put Harvey on the Scent," *Journal of the History of Medicine* 12 (1957): 102–105; E. T. McMullen, "Anatomy of a Physiological Discovery: William Harvey and the Circulation of the Blood," *Journal of the Royal Society of Medicine* 88 (1955): 491–498; Walter Pagel, "William Harvey Revisited," *History of Science* 9 (1970): 1.

5. William Harvey, *De motu cordis*, trans. Robert Willis (London: Printed for the Sydenham Society, 1847), repr., Great Books of the Western World, Vol. 28 (Chicago: Encyclopaedia Britannica, 1952). The 1847 translation by Robert Willis is used in this book.

6. Walter Pagel "The Philosophy of Circles—Cesalpino—Harvey: A Penultimate Assessment," *Journal of the History of Medicine* 12 (1957): 140–157.

7. Harvey, *De motu cordis*, introduction.

8. Fuchs, *Mechanization of the Heart*, 47.

9. Fuchs, *Mechanization of the Heart*, 10–15.

10. Harvey, *De motu cordis*, chap. 9; F. T. Jevons, "Harvey's Quantitative Method," *Bulletin of the History of Medicine* 36 (1962): 462–467.

11. Harvey, *De motu cordis*, chap. 9.

12. Harvey, *De motu cordis*, chap. 8.

13. Fuchs, *Mechanization of the Heart*, 38.

14. Whitteridge, *William Harvey and the Circulation*, 114.

15. Excellent translations of *De motu cordis* include the ones by Chauncey Leake, Kenneth Franklin, Gweneth Whitteridge, Robert Willis, and Geoffrey Keynes, which are listed in earlier notes.

22. CLOSING THE RING

1. William Harvey, *De motu cordis*, trans. Robert Willis (London: Printed for the Sydenham Society, 1847), repr., Great Books of the Western World, Vol. 28 (Chicago: Encyclopaedia Britannica, 1952), chap. 1.

2. J. B. West, "Marcello Malpighi and the Discovery of the Pulmonary Capillaries and Alveoli," *American Journal of Physiology, Lung and Cell*

Molecular Physiology 304 (2013): L383–L390. A good biographical summary with quotations is offered in Michael Foster, *Lectures on the History of Physiology*, lecture 4:86–99.

3. Borelli's research is the subject of lecture 3 in Foster, *Lectures on the History of Physiology*.

4. Foster, *Lectures on the History of Physiology*, lecture 4:86–99; Marcello Malpighi, *De pulmonibus, epistle I*, trans. N. J. De Witt, quoted in James Young, "Malpighi's 'De Pulmonibus,'" *Proceedings of the Royal Society of Medicine* 23 (1929): 1–11; Foster, *Lectures on the History of Physiology*, lecture 4:86–99, quoted in John Fulton, *Selected Readings in the History of Physiology*, 61; Alfred Fishman and Dickinson W. Richards, *Circulation of the Blood*, 29; Mark Graubard, *Circulation and Respiration*, 177.

5. Young, "Malpighi's 'De Pulmonibus,'" 23, 7.

6. Foster, *Lectures on the History of Physiology*, lecture 4:99–100.

7. C. V. Weller, "Antony van Leeuwenhoek," *Annals of Internal Medicine* 6 (1932): 573–584.

8. Foster, *Lectures on the History of Physiology*, lecture 4:100–101.

23. AHEAD OF THE CURVE

1. A fine study of the history of blood transfusion is Holly Tucker, *Blood Work: A Tale of Medicine and Murder in the Scientific Revolution* (New York: W. W. Norton, 2011).

24. PRIDE AND PREJUDICE

1. A comprehensive assessment of the reception of Harvey's work, both favorable and hostile, is Roger French, *William Harvey's Natural Philosophy* (Cambridge: Cambridge University Press, 1994); see also Humphrey Rolleston, "The Reception of Harvey's Doctrine," in *Essays on the History of Medicine Presented to Karl Sudhoff*, ed. Charles Singer and Henry Sigerist (Oxford: Oxford University Press, 1924), 248–254.

2. H. Bayon, "William Harvey, Physician and Biologist: His Precursors, Opponents, and Successors," pt. 2, *Annals of Science* 3 (1938): 83.

3. H. P. Bayon, "Allusions to the 'Circulation' of the Blood in MSS. Anterior to *De motu cordis* 1628" (Section of the History of Medicine), *Proceedings of the Royal Society of Medicine* 32 (1939): 707–718.

4. A. G. Debus, *The English Paracelsians* (London: Watts History of Science Library, 1965).

5. J. Schouten, "Johannes Walaeus (1604–1649) and His Experiments on the Circulation of the Blood," *Journal of the History of Medicine and Allied Sciences* 29 (1974): 259–279.

6. Thomas Fuchs, *Mechanization of the Heart*, 168.

7. Fuchs, *Mechanization of the Heart*, 151.

8. E. Gotfredsen, "The Reception of Harvey's Doctrine in Denmark," *Acta Medica Scandinavica Supplementum* 266 (1952): 75–85.

9. G. A. Lindeboom, "The Reception of Harvey's Theory of the Circulation of the Blood," *Janus* 46 (1957): 183–200.

10. C. Schmitt and C. Webster, "Marco Aurelio Severino and His Relationship to William Harvey: Some Preliminary Considerations," in *Science, Medicine, and Society in the Renaissance: Essays to Honor Walter Pagel*, Vol. 2, ed. Allen G. Debus (New York: Science History Publications, 1972), 63–72; C. Schmitt and C. Webster, "Harvey and M. A. Severino: A Neglected Medical Relationship," *Bulletin of the History of Medicine* 45 (1971): 49–75.

11. Gotfredsen, "The Reception of Harvey's Doctrine in Denmark," 202–208.

25. THE FRENCH CONNECTION

1. G. K. Talmadge, "Pierre Gassendi and the *Elegans de septo cordis pervio observatio*," *Bulletin of the History of Medicine* 7 (1939): 429–457.

2. René Descartes, *Discourse on the Method*, repr., Great Books of the Western World, Vol. 31 (Chicago, Encyclopaedia Britannica, 1952).

3. Rudolph E. Siegel, "Why Galen and Harvey Did Not Compare the Heart to a Pump," *American Journal of Cardiology* 20 (1967): 117.

4. Gweneth Whitteridge, *William Harvey and the Circulation*, 171–172.

5. William Harvey, *De motu locali animalium* (Local movements of animals), ed. and trans. Gweneth Whitteridge (Cambridge: Cambridge University Press, 1959).

6. R. B. Carter, *Descartes' Medical Philosophy: The Organic Solution to the Mind-Body Problem* (Baltimore: Johns Hopkins University Press, 1983).

7. Quoted in Louis Chauvois, *William Harvey: His Life and Times*, 188.

8. Thomas Fuchs, *Mechanization of the Heart*, 130.

9. A discussion of Descartes's views is in Michael Foster, *Lectures on the History of Physiology*, lecture 2:58–62.

10. Fuchs, *Mechanization of the Heart*, 128.

11. G. Gorham, "Mind-Body Dualism and the Harvey–Descartes Controversy," *Journal of the History of Ideas* 55 (1994): 211.

26. BAD BLOOD

1. Gweneth Whitteridge, *William Harvey and the Circulation*, 180–198.

2. William Harvey, "A Second Disquisition to John Riolan," in *The Works of William Harvey*, trans. Robert Willis (London: Printed for the Sydenham Society, 1847), repr., Great Books of the Western World, Vol. 28 (Chicago: Encyclopaedia Britannica, 1952).

3. William Harvey, "The First Anatomical Disquisition on the Circulation of the Blood, Addressed to John Riolan," in *The Works of William Harvey*, trans. Robert Willis.

4. Whitteridge, *William Harvey and the Circulation*, 185.

27. A CONFEDERACY OF *CIRCULATEURS*

1. G. A. Lindeboom, "The Reception of Harvey's Theory of the Circulation of the Blood," 183–200.

2. W. R. Lefanu, "Jean Martet, a French Follower of Harvey," in *Science, Medicine, and History, Essays on the Evolution of Scientific Thought and Medical Practice Written in Honour of Charles Singer*, Vol. 2, ed. E. Ashworth Underwood (Oxford: Oxford University Press, 1953), 33–40.

3. Quoted in Louis Chauvois, *William Harvey: His Life and Times*, 236.

28. ONCE MORE INTO THE BREACH

1. J. P. Arcieri, *The Circulation of the Blood and Andrea Cesalpino*; S. Peller, "Harvey's and Cesalpino's Role in the History of Medicine," *Bulletin of the History of Medicine* 23 (1949): 213–235.

2. Michael Foster, *Lectures on the History of Physiology*, lecture 2:34.

3. Foster, *Lectures on the History of Physiology*, lecture 2:34.

4. J. M. L. Piñero, "Harvey's Doctrine of the Circulation of the Blood in Seventeenth-Century Spain," *Journal of the History of Medicine* 28 (1973): 230–242; J. J. Izquierdo, "On Spanish Neglect of Harvey's 'De Motu Cordis' for Three Centuries, and How It Was Finally Made Known to Spain and Spanish-Speaking Countries," *Journal of the History of Medicine* 3 (1948): 105–124.

29. THE GOODNESS OF AIRS

1. Louis Trenchard More, *The Life and Works of the Honourable Robert Boyle* (Oxford: Oxford University Press, 1944).

2. Steven Shapin and Simon Schaffer, *Leviathan and the Air-Pump* (Princeton, NJ: Princeton University Press, 2017).

3. The best modern biography is Lisa Jardine, *The Curious Life of Robert Hooke: The Man Who Measured London* (New York: Harper Perennial, 2005).

4. Robert Hooke, "An Account of an Experiment Made by M. Hook, of Preserving Animals Alive by Blowing Air Through Their Lungs with Bellows," *Philosophical Transactions of the Royal Society of London* 2 (1667): 539, reproduced in John Farquhar Fulton, *Selected Readings in the History of Physiology*, 109, and in Mark Graubard, *Circulation and Respiration*, 207.

30. OXFORD CHEMISTS

1. Robert G. Frank, *Harvey and the Oxford Physiologists: A Study of Scientific Ideas and Social Interaction* (Berkeley: University of California Press, 1980).

2. Ebbe C. Hoff and Phebe M. Hoff, "The Life and Times of Richard Lower, Physiologist and Physician," *Bulletin of the Institute of the History of Medicine* 4 (1936): 517–535.

3. Richard Lower, *A Treatise on the Heart: Tractatus de corde*, with an introduction and translation by K. J. Franklin, Classics of Medicine Library (Birmingham, AL: Gryphon Editions, 1989). Extract quoted from K. J. Franklin's translation reproduced in Alfred Fishman and Dickinson W. Richards, *Circulation of the Blood*, 34–37.

31. NITER, NITER, EVERYWHERE

1. John Mayow, *Medico-Physical Works, Being a Translation of "Tractatus Quinque Medico-Physici, 1674,"* Alembic Club Reprints, No. 17, 2nd ed. (Edinburgh: Alembic Club, 1957), 73. Also quoted in Michael Foster, *Lectures on the History of Physiology*, 185–199; Mark Graubard, *Circulation and Respiration*, 229; and John Farquhar Fulton, *Selected Readings in the History of Physiology*, 111.

32. PHLOGISTON

1. A discussion of Black's research is in Michael Foster, *Lectures on the History of Physiology*, 232–236.

2. Extracts quoted in Foster, *Lectures on the History of Physiology*, lecture 9.

33. CONCERTO OF AIRS

1. Two popular biographies are W. R. Aykroyd, *Three Philosophers (Lavoisier, Priestley, and Cavendish)* (London: William Heinemann Medical Books, 1935); and Steven Johnson, *The Invention of Air: A Story of Science, Faith, Revolution, and the Birth of America* (New York: Riverhead Books, 2009).

2. Extracts quoted in Michael Foster, *Lectures on the History of Physiology*, lecture 9; John Farquhar Fulton, *Selected Readings in the History of Physiology*, 119.

34. UNTO THIS LAST

1. The definitive biography is Jean-Pierre Poirier, *Lavoisier: Chemist, Biologist, Economist*, trans. Rebecca Balinski (Philadelphia: University of Pennsylvania Press, 1996). See also Sidney J. French, *Torch and Crucible: The Life and Death of Antoine Lavoisier* (Princeton, NJ: Princeton University Press, 1941).

2. Another fine biography is Madison Smartt Bell, *Lavoisier in the Year One: The Birth of a New Science in an Age of Revolution* (New York: Atlas / W. W. Norton, 2005). Lavoisier's research in pneumatic chemistry is discussed in detail.

3. Extracts quoted in Michael Foster, *Lectures on the History of Physiology*, lecture 9; John Farquhar Fulton, *Selected Readings in the History of Physiology*, 122; see also Alfred Fishman and Dickinson W. Richards, *Circulation of the Blood*, 44–48.

4. Bell, *Lavoisier in the Year One*, 102–103.

5. Another splendid book on Lavoisier is F. L. Holmes, *Lavoisier and the Chemistry of Life: An Exploration of Scientific Creativity* (Madison: University of Wisconsin Press, 1987).

35. IN OUR TIME

1. Excellent personal explorations by the pioneers who developed modern techniques as applied to the understanding of contemporary cardiology are edited by Richard J. Bing, *Cardiology: The Evolution of the Science and the Art*, 2nd ed. (New Brunswick, NJ: Rutgers University Press, 1999); a partly autobiographical overview of the history of cardiology with a greater focus on contemporary heart disease is Sandeep Jauhar, *The Heart: A History* (New York:

Picador, 2019). Modern cardiology began in the 1940s, and the biography of America's greatest living cardiologist is essentially a biography of modern cardiology: Thomas Lee, *Eugene Braunwald and the Rise of Modern Medicine* (Cambridge, MA: Harvard University Press, 2013).

2. Michael Foster, *Lectures on the History of Physiology*, lecture 3.

3. Stephen Hales, *Statical Essays; Containing Haemastaticks, or An Account of Some Hydraulick and Hydrostatical Experiments Made on the Blood and Blood Vessels of Animals, etc.* (1733), repr., Classics of Medicine Library (Birmingham, AL: Gryphon Editions, 1987); extracts quoted also in Foster, *Lectures on the History of Physiology*, 231; John Farquhar Fulton, *Selected Readings in the History of Physiology,* 57; and Alfred Fishman and Dickinson W. Richards, *Circulation of the Blood*, 489.

4. Excellent sources on the history of hypertension are Fishman and Richards, *Circulation of the Blood*, chap. 8; and Arthur Ruskin, *Classics in Arterial Hypertension* (Springfield, IL: Charles C. Thomas, 1956).

5. John Floyer, *The Physician's Pulse-Watch; or, An Essay to Explain the Old Art of Feeling the Pulse, and to Improve It by the Help of a Pulse-Watch. In Three Parts* (1707), repr., Classics of Cardiology Library (Delanco, NJ: Gryphon Editions, 2018).

6. Foster, *Lectures on the History of Physiology*, lecture 3.

7. Bing, *Cardiology*.

8. André Cournand, with Michael Meyer, *From Roots . . . to Late Budding: The Intellectual Adventures of a Medical Student* (New York: Gardner Press, 1986).

9. Fishman and Dickinson, *Circulation of the Blood*, 50–54.

10. Lazzaro Spallanzani, *Experiments upon the Circulation of the Blood.* Reprinted in Classics of Cardiology Library (Delanco, NJ: Gryphon Editions, 2002).

11. Arnold Katz and Marvin Konstam, *Heart Failure: Pathophysiology, Molecular Biology, and Clinical Management*, 2nd ed. (Philadelphia: Wolters Kluwer/Lippincott Williams & Wilkins, 2009).

12. Ernest Baldwin, *Dynamic Aspects of Biochemistry*, 3rd ed. (Cambridge: Cambridge University Press, 1959).

13. Katz and Konstam, *Heart Failure*.

14. Hales, *Statical Essays*.

15. Auguste Krogh, "Reminiscences of Work on Capillary Circulation," *Isis* 41 (1950): 14. See also his Yale Silliman Lectures, summarized in A. Krogh, *The Anatomy and Physiology of Capillaries* (New Haven, CT: Yale University Press, 1922).

16. Fulton, *Selected Readings in the History of Physiology*, lecture 3; Fishman and Richards, *Circulation of the Blood*, chap. 6.

17. E. A. Shafer, *Textbook of Physiology* (Edinburgh, 1898), 1:285. See also his footnote 186.

18. Scott Hershberger, "Sketchable Sensor," *Scientific American*, November 2020, 19; "Big Questions from Ondrej Krivanek," *Scientific American*, November 2020, 3.

Index

LEEYA MEHTA

DHUN SETHNA, MD, is a clinical and academic cardiologist who has served on the senior academic staff at major medical centers, including the Cleveland Clinic, Cedars-Sinai Medical Center, and the Carilion Clinic. He has contributed to *Braunwald's Heart Disease: A Textbook of Cardiovascular Medicine*. He lives in Virginia.